W0112415

Cellular Development

OUTLINE STUDIES IN BIOLOGY

Editors : Professor T.W. Goodwin, F.R.S., University of Liverpool
Dr J.M. Ashworth, University of Leicester

Editors' Foreword

The student of biological science in his final years as an undergraduate and his first years as a postgraduate is expected to gain some familiarity with current research at the frontiers of his discipline. New research work is published in a perplexing diversity of publications and is inevitably concerned with the minutiae of the subject. The sheer number of research journals and papers also causes confusion and difficulties of assimilation. Review articles usually presuppose a background knowledge of the field and are inevitably rather restricted in scope. There is thus the need for short but authoritative introductions to those areas of modern biological research which are either not dealt with in standard introductory textbooks or are not dealt with in sufficient detail to enable the student to go on from them to read scholarly reviews with profit. This series of books is designed to satisfy this need.

The authors have been asked to produce a brief outline of their subject assuming that their readers will have read and remembered much of a standard introductory textbook of biology. This outline then sets out to provide by building on this basis, the conceptual framework within which modern research work is progressing and aims to give the reader an indication of the problems, both conceptual and practical, which must be overcome if progress is to be maintained. We hope that students will go on to read the more detailed reviews and articles to which reference is made with a greater insight and understanding of how they fit into the overall scheme of modern research effort and may thus be helped to choose where to make their own contribution to this effort.

These books are guidebooks, not textbooks. Modern research pays scant regard for the academic divisions into which biological teaching and introductory textbooks must, to a certain extent, be divided. We have thus concentrated in this series on providing guides to those areas which fall between, or which involve, several different academic disciplines. It is here that the gap between the textbook and the research paper is widest and where the need for guidance is greatest. In so doing we hope to have extended or supplemented but not supplanted main texts and to have given students assistance in seeing how modern biological research is progressing while at the same time providing a foundation for self help in the achievement of successful examination results.

Cellular Development

D.R. Garrod

Lecturer in Cell Biology
at Southampton University

Springer-Verlag Berlin Heidelberg GmbH

ISBN 978-0-412-11410-6   ISBN 978-1-4899-3374-4 (eBook)
DOI 10.1007/978-1-4899-3374-4
© *1973 D.R. Garrod*
Originally published by Chapman and Hall in 1973.
SBN 412 11410 0

This paperback edition is sold subject to the condition that it shall not, by way of trade or otherwise, be lent, re-sold, hired out, or otherwise circulated without the publisher's prior consent in any form of binding or cover other than that in which it is published and without a similar condition including this condition being imposed on the subsequent purchaser.

All rights reserved. No part of this book may be reprinted, or reproduced or utilized in any form or by any electronic, mechanical or other means, now known or hereafter invented, including photocopying and recording, or in any information storage and retrieval system, without permission in writing from the Publisher.

Distributed in the U.S.A.
by Halsted Press, a Division
of John Wiley & Sons, Inc
New York

Contents

1 Introduction

1.1. Aspects of development.

If you have been fortunate enough to see a film of the development of any multicellular organism or, better still, to watch live embyros developing, the intricate beauty of the developmental process will not have escaped you: nor will its complexity. Apparent complexity, however, is no reason for despair when one begins to think in terms of analysing development. Rather, it is a stimulus to the first and most important analytical step, that of simplifying the problem by dividing it into aspects which can be meaningfully studied.

The most obvious way to divide development is on a chronological basis – to begin with fertilization and proceed through cleavage, blastulation and gastrulation to organ fromation. Such a division is particularly useful for a descriptive study, but has the demerit that it tends to obscure similarities between different stages. Also, it is assumed that before reading this book you will have gained some knowledge of descriptive embryology. Our division will be, therefore, a mechanistic one which, in a sense, is both arbitrary and artificial, but which may be helpful in enabling the problems of development to be defined more clearly.

The development of multicellular organisms may be divided into three aspects, as follows.

1.1.1. Differentiation

This involves the structural and functional specialization of individual cells from one of a number of common basic cell types which are usually competent to develop in several different ways. Thus the mesenchyme cells of the embryonic chick limb bud may become, among other things, muscle or cartilage cells. Differentiation is largely an intracellular process involving the appearance of cells with certain biochemically or cytologically recognizable characteristics through the differential activation of genes whose products confer these characteristics on the cell. In skeletal muscle cells for example, specific proteins (actin and myosin) are synthesized, and arranged to give the typical striated appearance (Fig. 1.1a). (Differentiation is the subject of another book in this series, 'Cell Differentiation' by J.M. Ashworth.)

Recent advances in molecular biology have greatly stimulated research into differentiation and biochemical aspects of development. However, it is almost certainly mistaken to believe that biochemical dissection of the embryo will lead to a complete understanding of development, because biochemical studies in general throw little light on an equally fundamental aspect, cellular interaction. The two remaining aspects of our division of development are, first and foremost, problems of cellular interaction. These are the subjects of this book.

1.1.2. Pattern formation.

This is the spatial organization of differentiation In dealing with pattern formation, we are not concerned, *a priori*, with the intricate mechanisms by which individual cells change into, say, muscle cells or cartilage cells, but with the

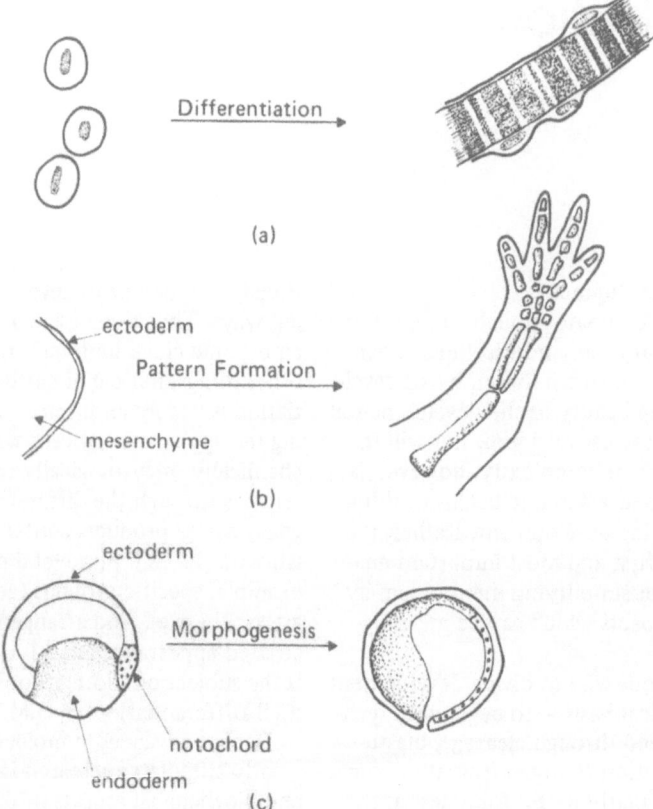

Fig. 1.1. (a) Differentiation: undifferentiated mesenchymal cells change into a muscle fibre. (b) Pattern formation: the limb bud gives an arm with characteristic patterned arrangement of skeletal elements. (c) Morphogenesis: Longitudinal sections of early and late amphibian gastrulae showing the alteration in arrangement of the germ layers.

development of muscle and cartilage in the correct spatial and proportional relationship to each other. The vertebrate limb begins development as a minute bud — a mass of undifferentiated mesenchymal cells covered by a single layer of ectodermal cells. From the bud arises a structure in which muscle, cartilage, nerves, blood vessels and skin are arranged in a very precise pattern (Fig. 1.1b). The problem is that of how this pattern is achieved.

1.1.3. Morphogenesis.

This is the mechanical process by which the form of the organism and the arrangement of its tissue are generated. Among other things, morphogenesis involves the co-ordinated movement of cells, sometimes as individuals, sometimes in large groups, the most dramatic example in most organisms being gastrulation. In amphibian development, a hollow ball of cells, the blastula, arises from the egg by a process of cell division

8

or cleavage. The three germ layers, the ectoderm, mesoderm and endoderm, are on the surface of the blastula, (Fig. 1.1c). During gastrulation the mesoderm and endoderm move inside, while the ectoderm spreads to cover them. Obviously, the rearrangement of embryonic tissues at gastrulation must be very precisely co-ordinated both in space and time. The problem is that of how the movements occur and how they are co-ordinated.

1.2. Model systems.

Although this book is entitled 'Cellular Development', a number of the examples used to illustrate the problems of pattern formation and morphogenesis are not taken from embryology, *sensu stricto*. The reason for this is that our present level of understanding suggests, paradoxically, that the provisional answers to embryological questions may be found elsewhere than in the embryo. Thus there are essentially two broad reasons for studying regeneration in *Hydra,* (see Chapter 3). The first is to obtain a better understanding of regeneration in *Hydra.* The second is that by doing this, it may be possible to gain insight into the problem of pattern formation in general, that is in both development and regeneration.

The second reason gives rise to the concept of 'model systems'. When a developmental biologist embarks upon a research project in an attempt to answer questions of fundamental interest to development, his choice of organism with which to work may be influenced to some extent by its suitability for research on his particular problem. He may refer to the organism as a 'model system', tacitly hoping that his results will be of general applicability. In the final analysis, however, any hypothetical extrapolation from a 'model system' to development in general must be tested by experiment.

2 Pattern formation

2.1. An outline of the problem.

The problem concerning us here is the spatial organization of cellular differentiation. Generally, adult organisms develop from a single cell, the egg, development involving a vast increase in structural complexity. The adult consists of many millions of cells of numerous different types and, although no two adults are identical, the degree of similarity between them is remarkable. All the tissues which the cells comprise are present in the correct proportions and in the same spatial relationship to each other.

For example, consider the normally developed human arm. It is always attached to the shoulder and not to the belly or the back of the head, and it always has four parts, the upper arm, the forearm, the wrist and the hand, which are always arranged in this order. It has skin on the outside, then muscle, then bone in the middle, to say nothing of a complicated pattern of nerves, blood vessels and other tissues. It never has bone on the outside and skin in the middle. These elementary facts may appear so obvious as to be hardly worth mentioning, but what is not at all obvious is how the developmental process achieves such a precise arrangement, or pattern, of the many parts and tissues of the adult organism. The understanding of pattern formation remains one of the central problems in developmental biology.

The problem is not confined to development, however. Many adult organisms are capable of regenerating lost parts. If a newt loses a leg, a replacement is formed; if an earthworm loses its tail, it grows a new one. Usually the regenerated part is an exact replica of the part which was lost, and is in perfect proportion to the rest of the body. Thus, if a newt loses a finger, it regenerates a finger; but if its forelimb is severed at the elbow, it regenerates a forearm, a wrist, and a hand. The regenerate corresponds precisely to what was lost, no more and no less. How is such precision achieved?

Pattern formation in regeneration and development are different because, in the former, structures arise *de novo*, whereas, in the latter, they arise by imposition of more complex organization on already existing embryonic structures. Also, there are two types of regeneration. *Morphallaxis* refers to regeneration which occurs as a reorganization of existing bodily structures without the addition of new material. *Epimorphosis* involves growth and the addition of new material, as well as patterned differentiation. However, our present level of understanding suggests that there may be distinct similarities between the processes involved, which justifies our considering them together.

2.2. General and theoretical aspects of pattern formation.

Three features of living organisms have long been recognized as being important in the regulation of cellular pattern, as follows:

2.2.1. Polarity.

During regeneration of an organ, or organism, the original axes are generally preserved. For example, if a piece is cut from the gastric region of *Hydra,* it regenerates distal structures (hypostome and tentacles) at its distal end, and proximal structures (peduncle and basal disc) at its proximal end. (Fig. 2.1.). The axial organization of the regenerate is the same as in the original animal, so that regeneration is polarized. Further, polarity is a property of the piece removed and does not depend on the presence of distal or proximal structures for its expression.

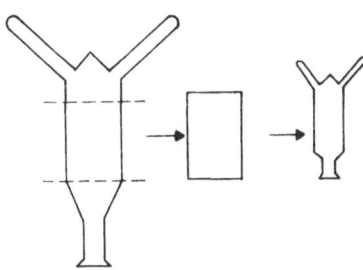

Fig. 2.1. Provided it is not too small, a piece cut from the gastric region of *Hydra* will regenerate in accordance with its original polarity. At first, the regenerate will be smaller than the original animal because regeneration in *Hydra* does not involve growth: it is an example of morphallaxis.

Polarity is also an important factor from the very earliest stages of development. In many cases, one or more of the axes of the future organism may be determined while the egg in still an oöcyte in the ovary.

2.2.2. Axial gradients.
Early experiments showed that certain properties were graded along the longitudinal axes of organisms capable of regeneration. If certain flatworms were cut transversely into pieces of equal size, heads were regenerated with greater frequency by the most anterior pieces, and the frequency of head regeneration declined in more posterior pieces. Antero-posterior gradients in

regeneration rate, metabolic activity and susceptibility to toxic chemicals were reported. Gradients of metabolic activity have also been discovered in embryos, for example the animal-vegetal gradient in sea urchin embryos [68].

2.2.3. Apical dominance.
During regeneration, the apical region, e.g. the hypostome in *Hydra* or the head in flatworms, is the first to form. Once formed, the apical region appeared to exert two important influences over the remainder of the regeneration process: first, to organize the properties of the regenerate, and second, to inhibit formation of further apical structures.

The discovery of axial gradients, and apical dominance, was an aspect in the formulation of the concept of a morphogenetic *field* which was defined by Huxley and deBeer [69] as 'a region throughout which some agency is at work in a co-ordinated way, resulting in the establishment of an equilibrium within the area of the field'. This definition was intended to distinguish a *field* system from a *mosaic* system as follows. In a mosaic system, removal of a part has no effect on the rest of the system, whereas removal of part of a field has an effect on the whole field and may result in proportionate regulation i.e. the reformation of a complete field. The field concept has given rise to some confusion and controversy which is discussed by Waddington [138]. Details of early work on pattern formation and the elaboration of the above concepts may be gleaned from the books by Huxley and de Beer [69] and Child [26].

Interest in the problem of pattern formation has revived during recent years, following a period in which biochemical and molecular aspects of development were more popular. Some new thinking about old concepts has taken place, and some new concepts have been added. In particular, there is the idea of *positional information* expounded by Wolpert [147, 148], which seems to give a new framework for looking at the problem. The suggestion is that cells may have their position specified within a field with respect to certain reference points. Each cell's genome might then respond to its positional information by instigating the appropriate type of molecular differentiation. In this way differentiation, a process which takes place within cells by the differential activation of genes and synthesis of specific proteins, could be spatially ordered to give rise to the overall cellular pattern of the organism.

Although a number of examples of pattern formation are given in the next chapter, it will be useful to consider at this stage one example which illustrates the interaction between the genome of cells and their position in the organism. Just below and behind the eye of most urodele tadpoles there is an ectodermal structure called the balancer (Fig. 2.2.). Although the balancer forms in this region, ectoderm from other parts of the embryo can form a balancer if transplanted to the same region. We say that ectoderm from other regions is *competent* to form balancer. If a piece of prospective balancer ectoderm is transplanted from one species to the same position in another species which does not normally have a balancer, it will differentiate to from balancer. The reverse experiment — transplantation of a piece of ectoderm from the region behind and below the eye of a species which does not normally form balancer to the same region in a species which does — results in absence of balancer in the tadpole. These experiments [82], (see [13]) suggest that the ectoderm responds to its position in the organism even if the organism is of another species, but the nature of the response depends on its genome, i.e. from which species it was derived.

Wolpert uses the French flag problem to illustrate his idea of positional information. Given a line of cells, each of which can differentiate to become either blue, white or red, how can they form a French flag which, from left to right, consists of equal blue, white and red

12

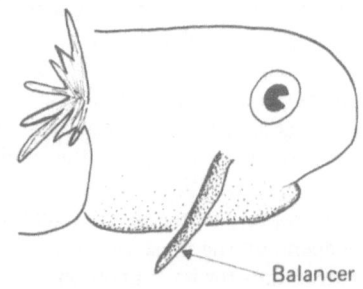

Fig. 2.2. Head of urodele tadpole showing balancer.

regions? The suggestion is that this could occur if the position of each cell were specified with respect to the ends of the line. For example, the ends could maintain different levels of a diffusible substance so that there would be a gradient of the substance along the line. The pattern could be formed if the cells responded by undergoing different types of differentiation at different levels of the gradient, that is at different concentrations of the gradient substance. In this illustration they would become blue at high levels, white at intermediate levels, and red at low levels (Fig. 2.3a.). Proportionate regulation could occur if the cut surface could reset the concentration of the substance at the same level as that which was maintained by the end removed (Fig. 2.3b.).

The idea of positional information and the French flag model are really statements of what a pattern-specifying mechanism has to achieve. The model makes use, in modified from, of the old concepts of axial gradients, polarity (the gradient slopes in a particular direction and the ends are different) and dominance (the ends set the level of the gradient). What is new is the suggestion of what type of information these factors, in combination, might give to the cells — they might enable each cell to 'know' its position within the system and differentiate accordingly.

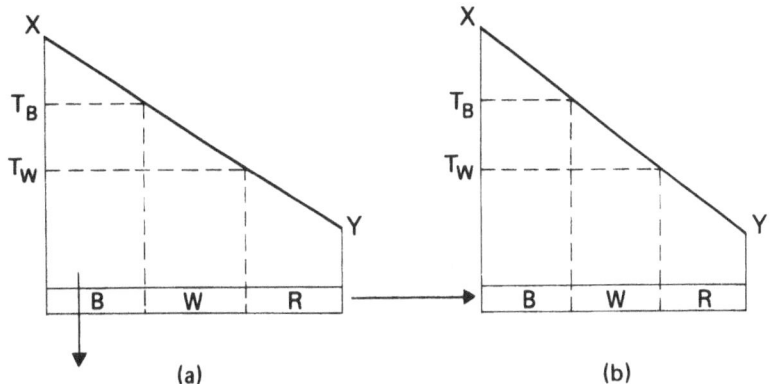

Fig. 2.3. The French flag model. (a) Generation of the pattern. Opposite ends of the line of cells maintain different levels, X and Y, of a diffusible substance so that a gradient is set up. Differentiation of the cells into blue, white or red depends on a threshold response to the concentration of the substance. If the level exceeds T_b, they become blue, if it is between T_b and T_w, they become white, otherwise they become red. (b) Regulation. If a cut is made at the arrow in (a) the cut surface re-establishes the level, X, of the substance. The gradient becomes steeper, but since the cells have the same threshold a proportionate, but smaller, French flag can be formed. (After Wolpert [148].)

In fact it appears that the simplest way of specifying positional information would be by means of an axial concentration gradient of a diffusible substance. Such a gradient could be set up in a number of ways. For example, the substance could be produced at one end of the system, and destroyed at the other. Wolpert has realized that morphogenetic fields are generally not more than 50 cells long (i.e. less than 2 mm) and suggests that the time required to signal position may be in the order of hours. Making certain assumptions about such properties as the diffusibility of the substance and membrane permeability, Crick [31] has demonstrated that it would be possible in principle to set up gradients of a few millimetres in this time.

Another interesting possibility for signalling position does not rely on a gradient. This is the phase-shift model of Goodwin and Cohen [54]. The phase angle difference between two wave-like propagations of biochemical activity, originating in a pacemaker cell (in the dominant region) and propagated from cell to cell, could specify position. This is analogous to measuring the distance from a thunder storm by the time interval between the flash of lightning and the clap of thunder.

Wolpert, and Goodwin and Cohen, give interpretations of various examples of pattern formation in terms of their respective ideas. We shall go on to consider specific examples of pattern formation in order to demonstrate the experimental approach to the problem. In some examples experimental results will be given with very little interpretation, while in others interpretations which can be made according to the results will be developed in some detail. Where interpretations have not been given, you may like to attempt your own: where they have been given, try to criticize them. An interesting game to play is to imagine that you are allowed to ask one question about each system, which has not already been answered by experiment. What question would you choose in order to

13

obtain the maximum amount of information about the mechanism of pattern formation? This is rather a diffcult game, however, so do not be discouraged if you do not succeed.

3 Specific examples of pattern formation

3.1. The cellular slime mould: a simple bi-partite pattern.

The slime mould species *Dictyostelium discoideum* has the simplest known cellular pattern. The pattern manifests itself during formation of the fruiting body which consists of only two cell types, spore and stalk. Fruiting bodies vary greatly in size, but spore and stalk cells are always formed in the same ratio, which is about 2:1 in the wild type: the pattern is size invariant [18]. The fruiting body is formed from a slug-shaped migrating mass of amoeboid cells, the grex or pseudoplasmodium. Although the grex appears homogeneous, it can be demonstrated by a number of techniques, immunological, histochemical, ultrastructural and behavioural,

that the pattern normally originates during grex migration (see [18]). The cells at the front of the grex become prestalk cells and those at the back, prespore cells. Even so, the migration stage is not essential for pattern formation since fruiting bodies may be formed directly from round aggregates of cells from an earlier stage in the life cycle [47], (Fig. 3.1.).

An experiment on the migrating grex by Raper [97] demonstrates two things: first, the pattern is capable of regulation, and second, that there is a difference between the front and back of the grex. The grex was divided transversely into four pieces of approximately equal size (Fig. 3.2.). Pieces from the back stopped migrating and, within a few hours, formed

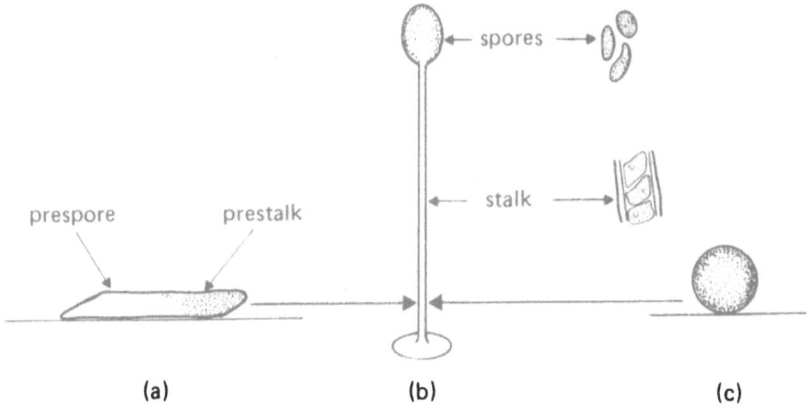

Fig. 3.1. Pattern formation in *Dictyostelium discoideum*. The fruiting body (b) consists of two cell types, spores and stalk (enlarged at right). It can be formed from the migrating grex (a) which contains prestalk and prespore cells, or directly from an aggregate of single cells (c).

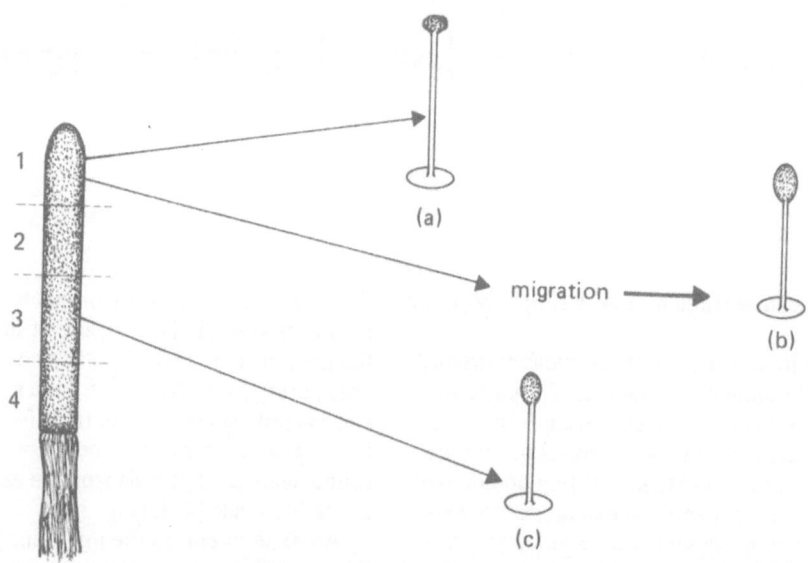

Fig. 3.2. Raper's experiment. The migrating grex was divided into four parts (numbered 1 to 4 from anterior to posterior). If piece 1 was allowed to form a fruiting body immediately, this consisted mainly of stalk (a), but fruiting body formation after migration for 24 h resulted in a return to more normal proportions (b). Piece 3 could form a fruiting body of normal proportions directly (c). (After Raper [97].)

fruiting bodies with normal proportions of stalk and spore, i.e. the pattern had regulated. The piece from the grex tip could be made to begin fruiting body formation immediately, or allowed to continue migration. In the former case an abnormal fruiting body was made, consisting mostly, or entirely, of stalk. If migration continued, however, the proportions regulated gradually, so that after about 24 h the tip formed a fruiting body with more normal proportions. It appears that the prespore cells at the back can regulate to form stalk fairly quickly, but that prestalk cells take rather longer to revert to spores.

What is the mechanism of pattern formation in the slime mould? The grex is formed by the chemotactic aggregation of single amoeboid cells. One view suggests that the two cell types may be present in the population even before aggregation and that these sort out in the grex to give the pattern [129, 21]. However, even if sorting out does take place, it seems unlikely to represent the whole story because the pattern regulates. Thus, if two cell types are present, they are not irreversibly determined as stalk and spore in the migrating grex, because cells of each type can form the other if the grex is cut. Presumably there is some pattern-determining mechanism within the grex, but very little, if anything, is known about it. A possible clue is given by a recent experiment in which slime mould cells were kept in a high concentration of cyclic-AMP [19]: they became stalk cells.

At the moment there is no coherent hypothesis relating to the mechanism of pattern formation in the slime mould. For students who wish to form their own opinions, Bonner's book [18] is recommended as a good review and source of references.

3.2. Hydra: a threshold-gradient model.

Contrary to previous views, the regeneration in *Hydra* appears to be an example of morphallaxis. There is no growth zone in the sub-hypostomal region and no apical growth takes place during regeneration [22, 27].

The hypostome is the dominant region in the classical sense. When grafted to the gastric region of another animal it organizes tentacle formation and the formation of a new axis. It also has an inhibitory action. When the hypostome and tentacles are cut off, they regenerate. However, if a hypostome is grafted laterally to the gastric region of a host, or onto the proximal end of the gastric region after removal of the peduncle, at the same time as the host's own hypostome is removed, distal regeneration of the host is inhibited [142].

Two interesting experiments suggested a model which could give regulation in *Hydra* [142, 140]. When Hydra is cut transversely the distal end of the proximal fragment acquires hypostomal properties before there is any overt sign of hypostome formation. Equal sized pieces were cut from different levels of the body and, after varying periods of isolation, grafted to the gastric region of host animals (Fig. 3.3.). If hypostomal properties had been acquired, a new axis would be organized in the host, otherwise the grafted piece would be absorbed. It was found that the time required for hypostome determination was graded disto-proximally, being much shorter in the sub-hypostomal region than in the peduncle (Fig. 3.4.). There was a gradient in time required for hypostome determination.

Next it was found that a sub-hypostomal region, removed from one animal and grafted directly into the gastric region of another, was absorbed, unless the host's hypostome was

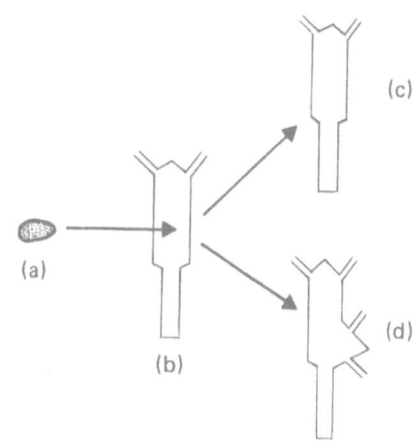

Fig. 3.3. The test for hypostome formation. An isolated fragment (a) was allowed to regenerate for a period of time and then grafted into the gastric region of a host Hydra (b). If the fragment had not acquired hypostomal properties it would be absorbed (c), but if it had, it would form a new axis and a new set of distal structures (d). (Partly after Webster [140].)

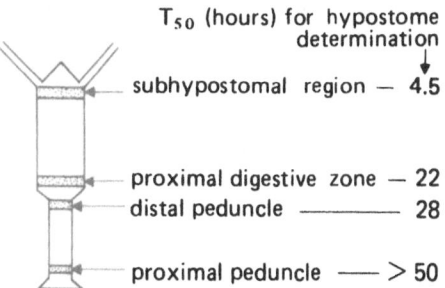

Fig. 3.4. The results of the experiment outlined in Fig. 3.3. The diagram on the left shows the regions of *Hydra* which were tested. The column on the right shows the T_{50} for hypostome formation in the respective regions i.e. the regeneration time necessary before half the isolated fragments would form a new axis in the host. (After Webster and Wolpert [142], and Webster [140].)

17

removed, in which case two axes regenerated. However, when a subhypostomal region was grafted to the basal disc of a host, it formed a second axis regardless of whether or not the host's hypostome had been cut off. Thus, although a hypostome inhibits the formation of another, the effectiveness of inhibition declines with distance from the hypostome.

The model suggested by these and other experiments is as follows. Suppose that the hypostome produces a substance which inhibits hypostome formation, and this substance is broken down elsewhere in the animal so that a disto-proximal gradient in inhibitor concentration is maintained. Suppose also that the cells have a threshold of response to the inhibitor which is also graded disto-proximally. When the level of inhibitor is above threshold, as is always the case in the intact animal, the cells cannot form a hypostome. When the level of inhibitor falls below threshold, as it would do if the hypostome were removed, hypostome formation begins. Cells at the distal end, able to form hypostome at a higher level of inhibitor than more proximal cells, would begin hypostome formation first, would recommence inhibitor production, and would inhibit further hypostome formation in proximal regions. Further evidence for the model is given by Webster [141].

This model is given in some detail because it illustrates the modern experimental and theoretical approach to the problem of pattern formation. However, it is by no means the only model for hydroid regeneration. An interesting alternative view has been taken by Rose [98]. Rose worked with the marine hydroid *Tubularia* which has a more complex structure than *Hydra*. His suggestion is that there is a 'hierarchy of self-limiting reactions' which means roughly that each region produces an inhibitor which prevents like differentiation in more proximal regions. Recently, some evidence supporting this view has been provided [99]. Factors have been isolated from various regions along the axis,

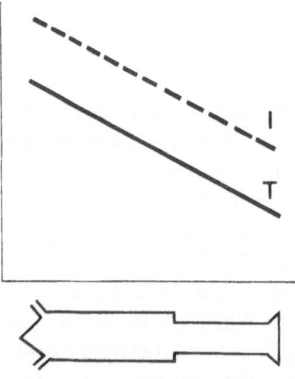

Fig. 3.5. The threshold gradient model. Webster suggested a disto-proximal gradient of an inhibitor substance (I) and a disto-proximal gradient in threshold for inhibition (T). The diagram suggests that the threshold in distal regions exceeds the inhibitor level proximally, so that a sub-hypostomal region grafted to the peduncle would not be inhibited from forming a hypostome. Substance I is supposed to be produced by the hypostome and destroyed elsewhere. Try to picture what would happen if the hypostome were cut off. (From Webster [140].)

which can inhibit distal regeneration in more proximal regions. These factors appear to be basic proteins.

The most recent work on *Hydra* has directed attention to polarity, and the importance of more proximal regions. Wilby and Webster [145], report that total polarity reversal can be brought about by grafting a hypostome onto the proximal end of the gastric region of an animal whose own hypostome has been removed (Fig. 3.6.). The sub-hypostomal region forms a peduncle instead of a hypostome. Faster polarity reversal occurs if, as well as the grafted hypostome, a peduncle is grafted with reversed polarity in place of the host hypostome. A peduncle thus grafted, can bring about partial polarity reversal by itself resulting in an animal with basal disc

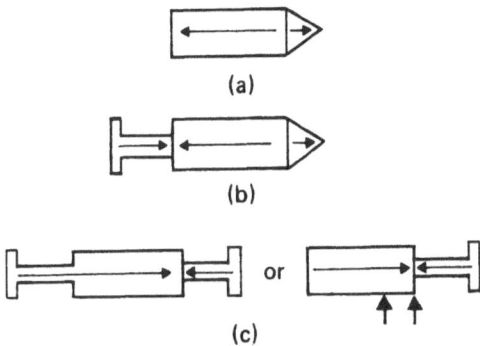

Fig. 3.6. Graft combinations which bring about polarity reversal in *Hydra*. The arrows indicate the original proximo-distal axis of the fragments. (a) Hypostome grafted to proximal end of gastric region. (b) Hypostome and peduncle grafted to gastric region. (c) Peduncle grafts. The small arrows in the right-hand of diagram (c) indicate alternative positions where lateral hypostomes were sometimes formed. (Partly after Wilby and Webster [145].)

and peduncle at both ends. These polarity reversal experiments often resulted in formation of a lateral hypostome which varied in position between the middle of the host's gastric region and the junction of the graft. The authors suggest that their results may be explained in terms of an antero-posterior gradient of a diffusible substance which is pumped (active transport) by the cells in a proximo-distal direction (for details see [145]). An interesting point from these experiments is that for the first time we find a hypostome forming in the middle of a gastric region in the absence of a cut surface, and at right angles to the main proximo-distal axis.

Wolpert, Hicklin and Hornbruch [150] point out that no case is known in which active transport of a substance *from cell to cell* takes place. They report in additional experiments on polarity reversal, and others emphasizing the importance of the basal disc or foot. If a peduncle is grafted to the gastric region of a host, it forms a foot, so that foot determination can be tested for in a similar way to hypostome formation [142]. It was found that the time required for foot determination decreased with increasing distance from the hypostome, taking 4 h at the bottom of the gastric region after removal of proximal structures, but 72 h in the sub-hypostomal region. The hypostome has an inhibitory effect on foot determination because, if the hypostome is removed as well as proximal regions, foot determination in the base of the gastric region takes only one hour, a remarkably short time. It is suggested that the foot is important either in fixing the level of one end of a disto-proximal gradient, or as a dominant region in its own right giving a substance which mirrors that at the head end.

As well as its obvious importance as an example of the modern approach to pattern formation, recent on *Hydra* is of interest in an historical context. Clearly, factors recognized by earlier workers, apical dominance, axial gradients and polarity, are still thought to be important, albeit in a somewhat modified form. The problem is to define these factors in biochemical and biophysical terms and to discover how they interact to produce proportionate regulation in the animal. Why, then, has recent work not adopted a more biochemical approach to the problem? One reason is that an understanding of properties and capabilities of the system, coupled with model building, may enable us to predict the type of biochemical mechanism involved (see [32]). Otherwise, looking for biochemical mechanism might be rather more difficult than looking for the proverbial needle in a haystack.

3.3. Limb development: ectodermal-mesenchymal interactions.

In the chick embryo, the wing first appears as a tiny bud about 3 days after the beginning of development (stage 16 of Hamburger and

19

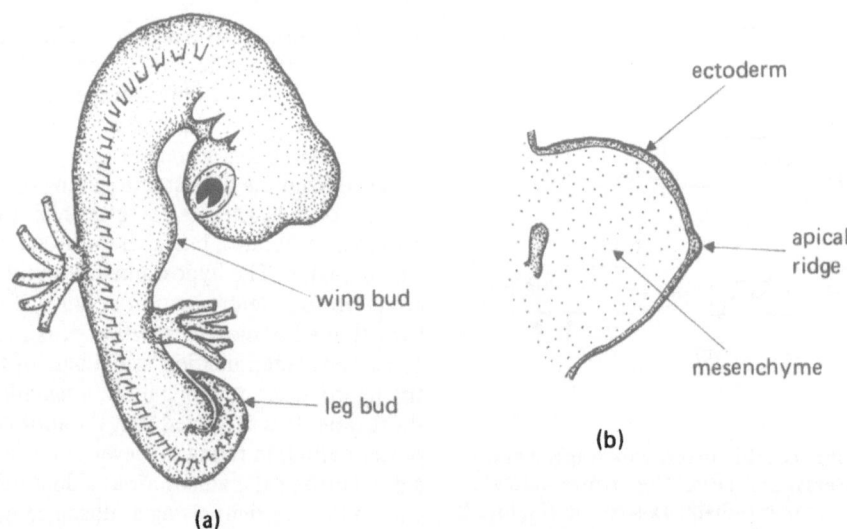

Fig. 3.7. Structure of the limb bud. (a) Chick embryo at stage 21 showing limb buds. (b) Section showing the ectodermal layer, the mesenchyme and the apical ridge. ((a) After Amprimo [9]; (b) after Saunders [105].)

Hamilton) (Fig. 3.7a.). The bud consists of a mesenchymal core and a one cell thick outer ectodermal layer. Soon after the appearance of the bud the distal ectoderm becomes thickened forming the apical ridge which runs antero-posteriorly around the bud margin (Fig. 3.7b.). The mesoderm and ectoderm, particularly the apical ridge, appear to have specific roles in limb development.

Saunders [105] showed that proximo-distal organization of limb structures took place in proximo-distal order in the limb bud between 3 and 4 days of development (stages 17 to 21 of Hamburger and Hamilton). That is prospective upper arm appeared first, followed by prospective lower arm and hand (Fig. 3.8.). However, the 4 day limb bud is not a mosaic because the parts are not irreversibly determined. Hampé [62] removed the middle piece of an early limb bud and grafted the tip onto the stump. Regulation took place and a normal limb bud resulted.

The importance of the apical ridge in proximo-distal organization was also shown by Saunders [105]. Removal of the ridge resulted in the absence of distal structures (Fig. 3.9.). The later the ridge removal, the more complete was the resulting limb. Subsequently, there was some disagreement about the importance of the apical ridge. The respective arguments are given by Zwilling [155], Amprino [9] and Faber [39]. Some of the points indicating the ridge's importance [155] are as follows. In wingless mutants, the ridge regresses very early in development of the bud. Grafting a second ridge onto a bud results in duplication of distal structures. Fragments of mesoderm, grafted to the chorio-allantoic membrane do not develop into recognizable limb structures unless they are enclosed in a jacket of ectoderm with an apical ridge. However, if limb mesoderm is grafted to the flank at a sufficiently early stage, the flank ectoderm forms a ridge and a wing is produced.

20

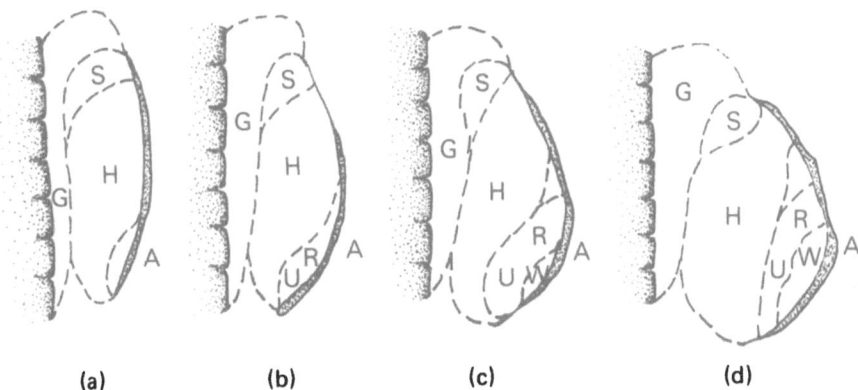

(a)   (b)   (c)   (d)

Fig. 3.8. Successive stages in wing bud development showing origin of prospective regions. The apical ridge (A) runs along the entire distal margin of the bud from anterior to posterior. G = shoulder girdle region; S = shoulder joint region; H = humerus region; R = radius region; U = ulna region; W = wrist and hand region (After Saunders [105].)

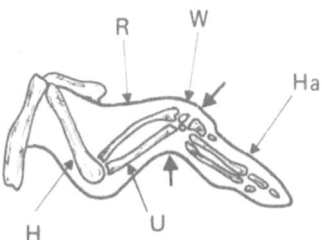

Fig. 3.9. Normal wing structure and effect of ridge removal. The diagram shows the outline of the wing and arrangement of the wing bones as they normally are from about 11 days of development. If the apical ridge of the wing bud is removed at about stage 21 (approximately 4 days) the structures distal to the arrows do not form. Lettering as in Fig. 3.8; Ha = hand. (After Saunders [105].)

Particularly interesting in relation to the role of the apical ridge are experiments in which the tip of the bud is removed, rotated about its disto-proximal axis and replaced. Rotation of the tip through 90° results in a 90° rotation of the distal parts of the final limb. Rotation through 180° results in duplication of distal structures, the two sets being mirror images of each other, i.e. the anterior set of distal structures are like those of the normal wing on the other side of the body (Fig. 3.10.) [106].

The pattern of axial structures can also be affected by transplantation of a small region of the mesoderm called the posterior necrotic zone (PNZ). This lies at the posterior margin of the limb bud, near its junction with the body and is called the PNZ because it is an area of extensive cell death. Transplantation of the PNZ to the distal tip of the limb bud again results in duplication of distal structures, though in this case they are not mirror images. Instead both sets of distal structure have the same antero-posterior organization which is appropriate to the limb on which the operation was done. Duplication with mirror imaging results if the PNZ is transplanted to the anterior margin of the wing bud. The result of this experiment is strikingly similar to that obtained when the tip of the bud is rotated through 180°.(Fig. 3.12b).

21

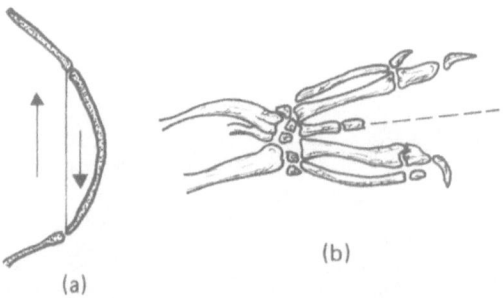

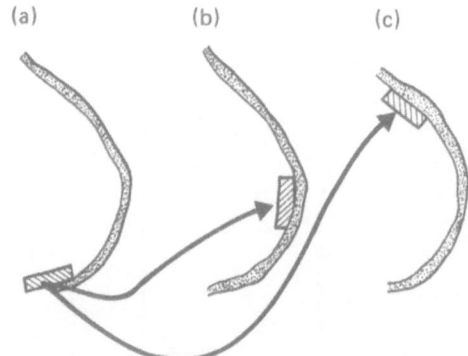

Fig. 3.10. Rotation of the apical region through 180°. (a) The experiment. The arrows indicate the posteri-anterior axes of the stump and the rotated tip. (b) The result. Distal structures are duplicated and are mirror images. The dotted line indicates the plane of symmetry. (After Saunders and Gasseling [106].)

Fig. 3.11. Transplantation of the posterior necrotic zone (PNZ). (a) The PNZ's normal location. (b) Transplantation to the bud apex. (c) Transplantation to the anterior margin of the bud. (After Saunders and Gasseling [106].)

The role of the PNZ in normal development is not clear because, if it is removed from the early limb bud, a normal wing still forms [106].

We shall avoid any interpretation of the above observation concerning the pattern of axial structures in the limb bud (you are referred to the review by Saunders and Gasseling [106]) and go on to consider another role of the mesoderm in limb development. There is evidence that the mesoderm is entirely responsible for determining the specificity of the limb. Sengel [111] has made all possible combinations between ectoderm and mesoderm of duck and duck limbs. The results indicate that in every case the origin of the mesoderm determined what type of limb was produced. Thus, if duck and leg mesoderm was enclosed in chick wing ectoderm, the resulting limb was a leg with a webbed foot. If chick wing mesoderm was combined with duck leg ectoderm, a wing with chick feathers was obtained. This is even more surprising because structures such as feathers and foot scales are of entirely ectodermal origin.

Nevertheless, the type of ectodermal structures produced were in accordance with the origin of the mesoderm.

In talking about limb development, none of the classical properties such as axial gradients and apical dominance have been mentioned because in general, the experimental facts do not seem to lend themselves to interpretation in these terms. There is a superficial similarity between the duplication of distal structures which results when a second apical ridge is grafted onto a limb bud and the formation of a secondary axis by as grafted hypostome in *Hydra*. However, it may not be helpful to look at limb development in these terms. Clearly there are important interactions between the mesoderm and ectoderm. Also certain specialized regions, the apical ridge and the PNZ, appear to be important in the axial organization of the pattern of limb structures, though exactly how they are involved is not clear at present.

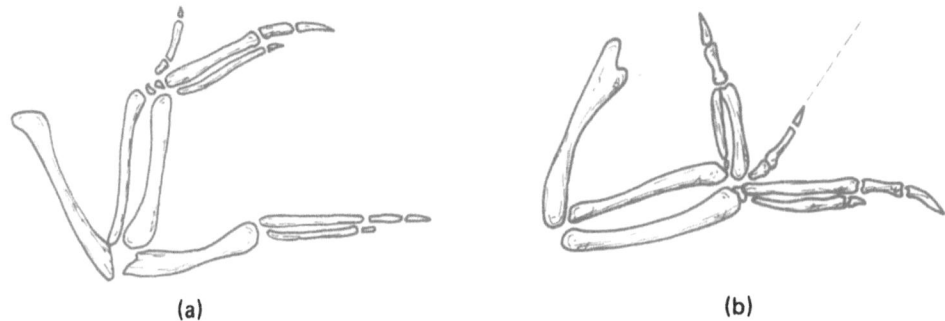

(a)   (b)

Fig. 3.12. Results of experiments indicated in Fig. 3.11. (a) Transplantation of PNZ to apical region of bud results in duplication of distal structures but not mirror imaging. (b) PNZ to anterior margin. The dotted line indicates the plane of symmetry. Note the striking similarity between this result and that obtained by rotating the bud through 180°. (Fig. 3.10.) (After Saunders and Gasseling [106].)

3.4. Limb regeneration: similarities between development and regeneration?

Here we are concerned with regeneration of limbs in adult amphibians, these being the most advanced vertebrates (in the evolutionary sense) in which complete limb regeneration occurs. This is an example of epimorphosis: new parts are regenerated from a mass of undifferentiated cells, the blastema, which forms at the cut end of the stump and grows during redifferentiation [63]. The sequence of events which takes place following amputation of a limb at any level is as follows (Fig. 3.13.). (a) Wound healing: the epidermal cells move to cover the exposed surface of the wound. (b) 'Dedifferentiation': appearance beneath the wound epithelium of cells which do not show cytological characteristics of differentiated cells such as cartilage and muscle, probably because they have lost these characteristics. (c) Blastema formation: organization of undifferentiated mesenchymal cells into a cone-shaped structure covered by an apically-thickened epidermal layer. (d) Redifferentiation and Morphogenesis: as the blastema grows, cells showing characteristics of muscle and cartilage reappear in the correct spatial relationship to each other and to the corresponding tissues in the stump.

The important problem is how the tissues which develop from the blastema are harmonized with those of the stump so that the new parts are exact replicas of those which were removed. We shall see later that the young blastema can develop autonomously if isolated, though in normal regeneration the influence of the stump is clearly important. The influence of the stump on regeneration has been the subject of much experimentation in which certain important factors have been identified. This work was the subject of a review by Goss [55] from which much of what follows has been taken and which gives the original references.

Denervated limbs do not regenerate, unless they have been allowed to develop without nerves in the first place (so-called aneurogenic limbs). However, the nerve requirement for limb regeneration does not appear to be specific. That is to say, in order to regenerate normally the limb stump does not seem to need an appropriate specific pattern of limb nerves, but rather a minimum number of nerves − a non-specific, threshold nerve supply. At present it seems that

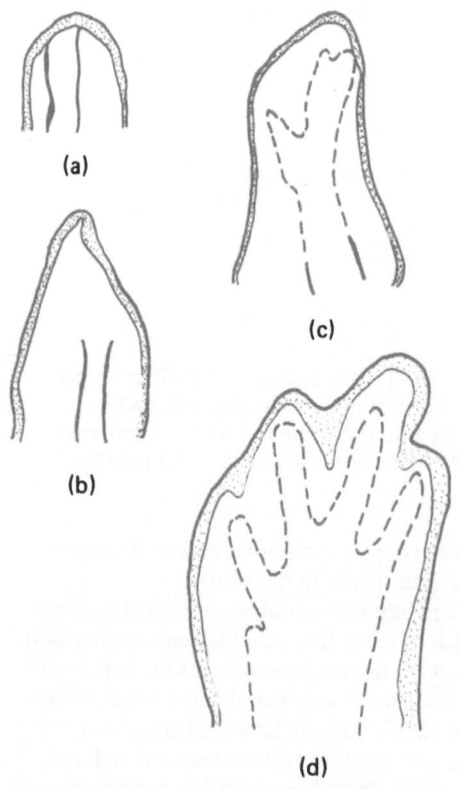

(a)

(b)

(c)

(d)

Fig. 3.13. Stages in the regeneration of salamander limb. (a) 3 days after cutting. The ectoderm has covered the wound surface. (b) Cone-shaped blastema. Note thickening of apical ectoderm. (c) Palette stage. Cartilaginous hand structures (dotted lines) beginning to form. (d) Five finger stage. Cartilaginous hand structures well defined. (From photographs of Mettetal [85].)

we may ignore nerves as a pattern forming factor in limb regeneration.

Up to a point the same appears true of bone in that total bone removal in the stump does not preclude normal regeneration. However, bones do appear to play an important role in axial pattern and size determination. Thus if

distal bones (radius and ulna) in the stump are replaced by proximal bones (humerus), the resulting regenerate is longer than expected, and shorter if proximal bones are replaced by distal. The regenerate appears to measure proximo-distal distance from the bone at the distal end of the stump, so that, in the case where amputation through the forearm was followed by replacing radius-ulna with humerus, the regenerate behaves roughly as though amputation had taken place through the upper arm. Interchanging humerus and femur has no effect on the size of the regenerate.

It is difficult to test for a requirement for muscle in regeneration. However, there is some evidence that muscle controls the specificity of limb regeneration as we have seen in the case of mesoderm in avian limb development. For example, replacement of axolotl limb stump muscle with tail muscle may result in a tail-like structure being regenerated from the limb stump.

Skin appears to be necessary for limb regeneration, though again, as in the case of avian limb ectoderm, it does not appear to control specificity. Limb regnerates may be obtained from limb stumps covered with tail ectoderm, unless the stump is irradiated prior to skin transplantation, in which case tail-like regeneration may occur.

Interesting results are obtained if two limb stumps, with their tips joined together, are allowed to regenerate side-by-side. If the two stumps are parallel, new distal structures are double. As the angle between the proximo-distal axes of the limb is increased, duplication is progressively reduced until, at a certain angle, the two stumps form and share a single set of distal structures. This suggests some kind of rigid proximo-distal influence within the stump, which is involved in anterio-posterior determination and which can interact with or cancel out a similar influence from another stump.

In some instances, proximo-distal polarity appears to be determined in relation to the rest

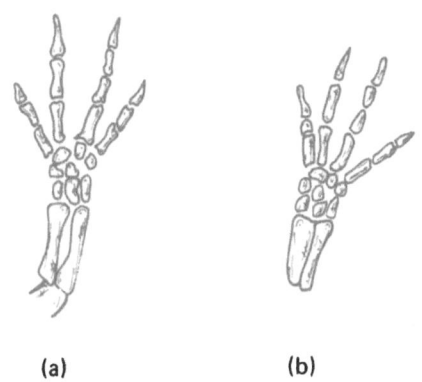

(a)   (b)

Fig. 3.14. Development of cone-shaped blastema transplanted to dorsal fin. (a) Normal *Ambystoma* fore-limb showing point of transection. (b) 25 day regenerate of upper arm blastema transplant. (From photographs of Stocum [124].)

of the body. Thus if a piece of limb is grafted to the body with reversed proximo-distal polarity, a limb with normal polarity with respect to the body is formed. However, if proximally-directed regeneration is allowed to occur from distal parts, e.g. by cutting transversely half way through the limb and allowing regeneration from both cut surfaces, mirror imaging of distal parts occurs.

Turning now to the blastema, it was believed at the time of Goss' review [55] that it had little or no autonomous organizing ability. Recent work by Stocum [123, 124] and de Both [38] has demonstrated, on the contrary that the isolated blastema is able to differentiate and form normally patterned limb structures. Firstly, Stocum [123] showed that very early blastemas of *Ambystoma,* isolated before any sign of differentiation was detectable and maintained in tissue culture, could form muscle and cartilage. Further, differentiation was poorer when part of the stump was cultured with the blastema

than when the blastema was cultured alone. Differentiation in tissue culture was not patterned, but when similar early blastemas were transplanted to the dorsal fin, patterned differentiation took place (Fig. 3.14.) although there was frequently loss or fusion of some skeletal elements [124]. Transplantation of blastema and stump resulted in normal regeneration. The blastemas used were obtained by amputating the arm through the distal humerus, so it is interesting that early distal half blastemas formed hand structures when transplanted. Early proximal half blastemas were either absorbed, or formed hand structures also, whereas slightly later proximal half blastemas formed lower arm structures.

De Both [38] transplanted early upper arm blastemas to the flank or orbit. Single transplanted blastemas formed forearm and hand structures, but when several similar blastemas were fused, upper arm structures were formed as well. The same was true of later, so-called paddle-shaped blastemas in which redifferentiation of skeletal structures had begun. One blastema gave hand structures, but several combined gave forearm structures in addition.

These results demonstrate several interesting points. First, the blastema is capable of autonomous, normal, patterned differentiation. Second, like the early limb bud, the blastema is clearly not a mosaic since fused blastemas from the same level can regulate to form proximal structures. Third, unlike the limb bud, it seems that distal tendencies appear first in the blastema, since early upper arm blastemas form hand structures. However, in spite of the autonomous capabilities of the blastema, the stump is obviously important in normal regeneration, because the blastema left in position forms only these structures which have been removed. Lastly, there is a possibility that the apical ectoderm of the blastema may have properties in common with the ectodermal ridge of the limb bud. Unfortunately for the experimentalist, the

25

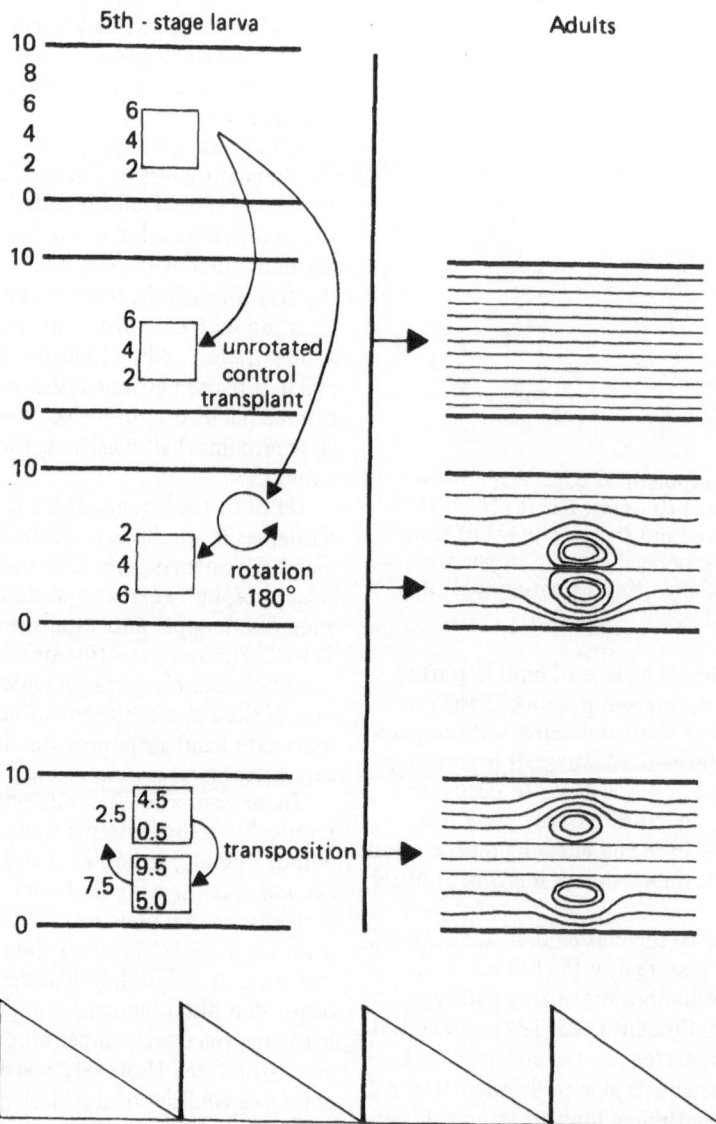

Fig. 3.15. Cuticle transplantation in *Rhodnius*. The experiments performed in the 5th-stage larva are shown on the left and the resulting adult ripple pattern on the right. The figures indicate the height of the gradient in the host segment and transplants. The diagram at the bottom shows the suggested serially repeated segmental gradients, the thick lines representing intersegmental margins. (After Locke [79]; from Lawrence [77].)

blastema ectoderm regenerates rapidly if it is removed, but by transplanting an apical cap to the base of another blastema some duplication of distal structures may be obtained [131].

As in the case of the limb bud, limb regeneration does not appear to lend itself to theorizing in terms of classical concepts. The finding that distal determination seems to occur first in the blastema, has something in common with the idea of apical dominance and the formation of the hypostome in *Hydra*. However, it does not necessarily follow that this is a helpful way of looking at the problem.

3.5. The insect cuticle: studies on polarity.

As an experimental system the insect cuticle has the advantage that the orientation of external structures indicates the polarity of the single layer of epidermal cells which lie beneath it. In the hemipteran *Rhodnius*, small cuticular ridges called ripples run transversely across the segment, at right angles to the polarity of epidermal cells, while in another bug, *Oncopeltus*, hairs point posteriorly, in line with the axis of polarity. Certain experiments performed during the last fifteen years by different workers, when looked at together, enable suggestions to be made about the nature of the polarity system of the cuticle, [77]. The experiments usually involved transplanting pieces of larval cuticle and observing the effect on the pattern in the adult after ecdysis.

The reason for suggesting a gradient in each segment came from the work of Locke [80]. Square pieces of *Rhodnius* cuticle were removed and either rotated through 90° or 180°, or transplanted to a different level in the segments, e.g. pieces from the anterior end of the segment were transplanted to the posterior end. Specific alterations in the ripple pattern resulted (Fig. 3.15.). If pieces were removed and latterally transplanted to the same level in the segment, or transplanted to the same level in another segment, the ripple pattern was unaltered.

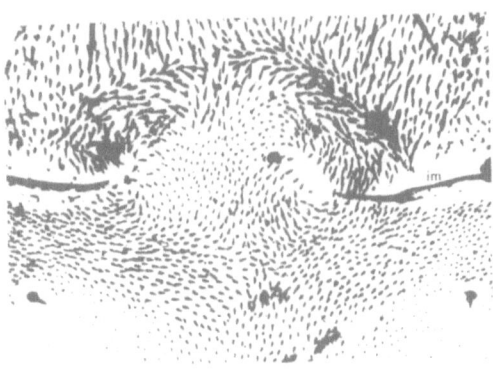

Fig. 3.16. The hair pattern found near a discontinuity in an intersegmented membrane in *Oncopeltus* (Drawn from Lawrence [77].)

It appeared that the gradient ran antero-posteriorly and was repeated in each segment.

Insect segments are separated from each other by intersegmental membranes which bound the gradient systems of adjacent segments. Lawrence [77] observed alterations in the cuticular hair pattern of *Oncopeltus* in insects which had gaps in the intersegmental membranes, (Fig. 3.16.). The modified patterns suggested certain properties of the gradient system. An analogy was drawn between the segmental gradient and the behaviour of a gradient of sand. In the sand model, the intersegmental membrane was represented by two glass plates (Fig. 3.17.). Two sand gradients were built up on opposite sides of the plate, with the high end of one gradient on one side and the low end of the other gradient on the other side. When the plates were separated, the sand flowed through the gap from the high end of the posterior gradient and formed a new stable gradient (the gradient is stable because equilibrium is achieved when the force of gravity and friction between the sand grains balance) which bore a striking resemblance to the cuticular gradient suggested by the

27

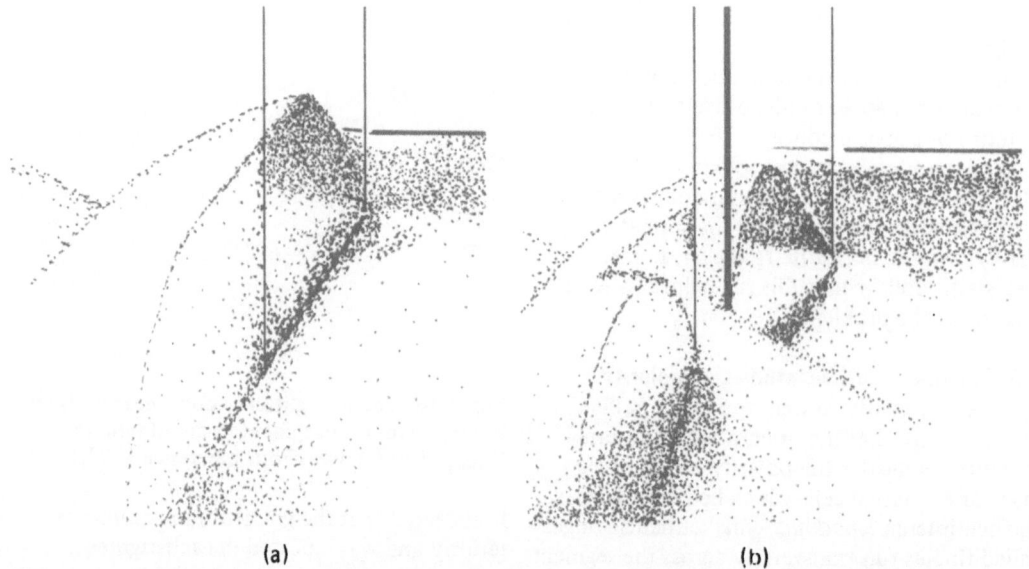

(a)   (b)

Fig. 3.17. The sand model. (a) Two glass plates separate two sand gradients. (b) Separation of the plates allows the sand to flow through. The contours of the new sand gradient are identical with that suggested by the orientation of hairs in Fig. 3.16. (Drawn from Lawrence [77].)

orientation of the hairs around the gap in the intersegmental membrane. It is possible to suggest that the segmental gradient is of a diffusible substance, the concentration of which decreases antero-posteriorly along the segment. When there is a gap in the intersegmental membrane, a new concentration gradient is present so that the substance diffuses anteriorly through the gap. In order that the resulting gradient should be stable, it is necessary to suppose that the epidermal cells transport the substance against its concentration gradient resulting in an equilibrium between diffusion down a concentration gradient (gravity in the sand model) and pumping against the gradient (friction between the sand grains). The difference between the sand model and the cuticular gradient is that the former is static and the latter dynamic.

28

An important question is whether the intersegmental membrane plays an active role in generating the gradient, or merely separates the gradients in adjacent segments. For example, the anterior and posterior margins of the intersegmental membrane could maintain different levels of the diffusible substance [126]. In the wax-moth, *Galleria*, at least, there is evidence that the intersegmental membrane plays an active role. Implantation of pieces of segment margin into a segment altered the cuticular scale pattern such that their orientation appeared to be influenced by both the implanted piece and the normal segment margin [83, 126]. However, in *Galleria* the anterior and posterior segment margins are different and it is possible to show that both are not essential for scale orientation. Piepho [95] produced, by grafting, a piece

Fig. 3.18. Piece of *Galleria* cuticle entirely surrounded by posterior segment margin showing centripetal alignment of scales. (Drawn from Piepho [95].)

of adult cuticle entirely surrounded by posterior margin (Fig. 3.18.). The scales within this region were centripetally orientated suggesting a radial gradient. Such a gradient could arise if, for example, the diffusible substance were produced by the segmental margin and were either destroyed by the epidermal cells or broke down spontaneously.

More recent work concerned the ripple pattern obtained when a piece of *Rhodnius* cuticle is rotated through 90° [78]. Making certain assumptions about the supposed gradient, it was possible to compute what the ripple pattern resulting from such a rotation should be. If it was assumed that, instead of pumping against the gradient, the epidermal cells 'remembered' their individual levels of the gradient substance and attempted to maintain that level after the operation, remarkable similarity between the observed and expected results was found.

It should be realized that the ability of a model to predict an experimental result does not necessarily prove that the model is correct because there may be other ways of achieving the same result. The value of the type of thinking which has been applied to work on the insect

cuticle, and of models in general, is that they point to questions which may be asked experimentally and which, when answered, may help us to understand the properties of the system.

3.6. The formation of nerve connections: a different type of problem?

Because of its tremendous complexity, the nervous system may seem forbidding from the point of view of studying pattern formation. It does not necessarily follow, however, that the mechanism of pattern formation will turn out to be more complex in principle than in other overtly simpler examples, though it may well be more precise in detail.

The ability of organisms to make co-ordinated locomotory responses to environmental stimuli necessitates, among other things, a system of precisely patterned nerve connection between peripheral sense organs and the central nervous system, within the CNS, and between the CNS and the body muscles. The problem concerning us is how the pattern arises. In considering this we shall deal with only one aspect, the formation of neuromuscular connections, which will serve to illustrate the nature of the problem. This and other aspects have been discursively reviewed by Gaze [46] whose book is recommended to anyone who finds what follows interesting.

In vertebrates, the cell bodies of motor neurones are situated in the ventro-lateral region of the spinal cord. Axons grow out from the cell bodies to innervate the muscles. In amphibia and fishes axon outgrowth and muscle innervation occur during both development and regeneration, but there is a distinct difference between developmental innervation and, for example, the reinnervation of a transplanted limb. During amphibian development, nerve fibres grow into the limb bud before muscle formation has taken place, [130]. During reinnervation, on the other hand, the nerves go to muscles which are already formed, but even so normal co-ordination may be restored.

29

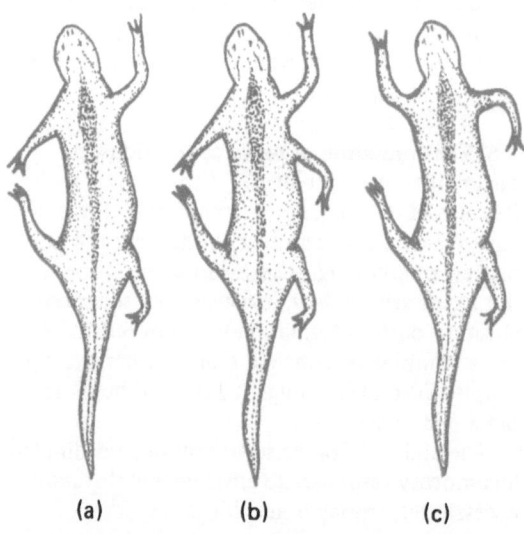

(a) (b) (c)

Fig. 3.19. The homologous response. (a) Normal newt. (b) Left fore-limb grafted to right hand side behind normal fore-limb. (c) Left and right fore-limbs interchanged.

An important example of reinnervation and restoration of function is the so-called 'homologous response' which comes from the early work of Weiss (see [46] for references). It was found that when a limb of a larval urodele was transplanted adjacent to the normal limb of a host animal, the transplant would eventually show the same motor activity as that of the normal limb, regardless of its orientation. Thus, if a left fore-limb was transplanted to the right side of the host just behind its own normal right fore-limb, it would eventually show responses which were mirror images of the responses of the host's limb (Fig. 3.19.). When the host limb moved forward, the transplant would move backwards, and so on. Although the transplanted limb was now on the right hand side of the body and although it moved simultaneously with the host's normal right limb, its movements

were appropriate to the left hand side of the body from which it came. The homologous response is also shown when right and left fore-limbs are interchanged in the same animal. When function is restored, each limb operates with a normal sequence of muscular activity, but because it is reversed, tends to move the animal backwards instead of forwards, and will do so if the normal hind limbs are paralysed.

To quote from Gaze [46], 'when the muscles in a transplanted limb show homologous response it appears that the central nervous system is calling them into action by name, so to speak. When the biceps of the normal limb contracts, so does the biceps in the transplant; when the triceps in the normal limb contracts, so does the triceps in the transplant. This is evidence that the central nervous system "knows" which muscle is which in the transplant.'

There are several possible explanations of how the CNS 'knows' which muscle is which, but recent evidence seems to favour the idea of 'selective reinnervation'. This suggests that the appropriate nerves growing out from the CNS somehow find the right muscles and form connections with them. For example, when nerves appropriate for biceps innervation grow into a transplanted limb, they find the biceps, irrespective of the orientation of the transplant and, therefore, of whether or not the biceps is in its normal position.

Two recent papers [127, 119] provide evidence in favour of this view. Székely showed that if a pair of forelimbs were transplanted to the thoracic region in a urodele larva, they did not become innervated by motor nerves. If, however, in addition to the limbs, the brachial region of spinal cord was also transplanted, the limbs acquired motor nerves and functioned normally (Fig. 3.20.). The brachial region of the cord is that from which motor innervation of the forelimbs normally comes, so the experiment suggests that only the appropriate nerves can innervate the limb muscles: thoracic motor

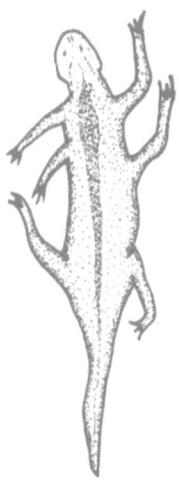

Fig. 3.20. Pair of fore-limbs transplanted to thoracic region together with section of brachial spinal cord. The limbs move in the same way as normal fore-limbs. (Drawn from Szekely [127].)

nerves would not do in place of brachial motor nerves.

Sperry and Arora worked on the occulomotor innervation of the eye muscle in fishes. They showed three things. (a) If the occulomotor nerve was cut and its various branches allowed to regenerate to the appropriate muscles, reinnervation and normal eye movements were rapidly restored. (b) If, after cutting, the occulomotor branches were prevented from reconnecting with the appropriate muscles, reinnervation by inappropriate nerves took place, but the resulting eye movements were weak and abnormal. (c) If innervation by inappropriate fibres was allowed to occur and then the appropriate nerves were allowed to regenerate, they competed with the inappropriate nerves, so that normal innervation and eye muscle function were restored.

Both the above pieces of work seem to support the idea of 'selective reinnervation' in favour of the main alternative hypothesis, 'myotypic respecification', which suggests that muscles become non-specifically reinnervated and that the nerves themselves, perhaps by way of their central connections, become respecified in relation to the muscles they have innervated.

If selective reinnervation does take place, what is its mechanism? How do the fibres find the right muscles? Some experiments suggest that ingrowing fibres do not need to be able to follow the pre-existing paths left by the original fibres which degenerate when a limb is transplanted. For example, aneurogenic limbs of *Ambystoma* (i.e. limbs which have developed without a nerve supply) were able to develop normal reinnervation patterns when transplanted to normal positions in normal larvae [92]. Later experiments showed that normal reinnervation did not occur if aneurogenic fore-limbs were transplanted in place of hind-limbs, instead of at the orthotopic site, [93, 94].

In conclusion, it must be stressed that the above presents only a tiny part of the published work on pattern formation in the nervous system and the arguments have been grossly simplified for brevity. A wider and more balanced understanding of the problems can only be achieved through wider reading, for which Gaze's review is an excellent starting point.

3.7. Primary embryonic induction.
Primary embryonic induction, the formation of nervous system from the ectoderm overlying the dorsal invaginated mesoderm in vertebrate embryos, is the most commonly described patternforming mechanism in text books of embryology, so it will not be considered in detail here. It has been thought for some time that induction involves the transfer of specific substances from the mesoderm to the neurectoderm, and that these substances 'evoke' the neuralizing response. There is some evidence for two proteinaceous inducer substances, the archencephalic or neuralizing inducer and the spinocordal

31

or mesodermalizing inducer, which are supposed to interact to organize the pattern of the nervous system and other axial structures (see [107, 132, 41]).

Recently, an interesting new idea has been suggested in this connection [30] based on the Goodwin—Cohen model (see Chapter 2). It is suggested that two pacemaker centres which originate propagated waves of biochemical activity may be present in the amphibian embryo. Before gastrulation, these may be localized at the animal pole (centre 1) and in the grey crescent region (centre 2). During gastrulation centre 1 moves towards the dorsal lip of the blastopore, while centre 2 invaginates and becomes stretched out longitudinally in the dorsal mesoderm. Cooke and Goodwin suggest that two such centres could give rise to a two-dimensional grid of positional information (Chapter 2) in the mesodermal mantle, which could specify the pattern of tissue differentiation.

4 Cell movement and its control in morphogenesis

Research in the field of morphogenesis is largely concerned with trying to understand the co-ordinated movements of cells and cell populations. For example, the dramatic morphogenetic changes which occur during amphibian gastrulation (see Chapter 1) are spatially and temporally co-ordinated movements of large groups of cells. The endoderm and mesoderm begin invagination at the end of blastulation and by the end of gastrulation, have moved from the surface of the embryo to the inside. The internal arrangement which they achieve is the basic plan for the tissue arrangements in the fully developed organism, and the correct spatial arrangement of tissues in the gastrula is crucial for the next stages of development.

Morphogenesis gives rise to two important questions which will be considered in some detail: (i) How do cells move? and (ii) How are their movements co-ordinated? Unfortunately it has been possible only in rare cases to make direct observations of cell movement in embryos because most embryos are not transparent, so that work relating to these questions has been done largely with cells in artificial situations such as tissue culture. For this reason we shall consider the problems of the mechanism of cell movement and its co-ordination in general, and then go on to deal with specific cases of morphogenesis in developing organisms in the next chapter.

4.1. The mechanism of cell movement.

It is possible to make two generalizations about the movement of embryonic cells. Firstly, these movements fall into the general class of 'amoeboid' movement. That is to say there is usually some kind of 'pseudopodal' activity at the front end of advancing cells, though different types of cells show specialized forms of this (Section 4.2). Secondly, it seems fairly clear that a cell moving over a substratum must adhere to the substratum and, further, must be able to make adhesions at its advancing end and break them at the back. Precisely how a cell's pseudopodal activity and adhesive properties are involved in bringing about movement over the substratum is debatable and leads to a principal controversy in the field of cell movement, that of how the driving force for movement is produced. Three main ideas have been put forward.

4.1.1. The cell membrane as the site of production of the force for cell movements.

Several workers have suggested that the force for cell movement may be produced by turnover of the surface membrane [14, 52, 116, 117]. The suggestion is that new cell surface is formed at the advancing end of a cell and that surface resorption takes place at the back (Fig. 4.1.). This implies that the lateral surface remains stationary relative to the substratum as the cell moves forward and that the whole surface is completely renewed each time a cell moves through its own length. This suggestion is based on the observation of particles attached to cell surfaces, which, in some cases, remain stationary relative to the substratum as the cell moves forward. It has been felt that this particle

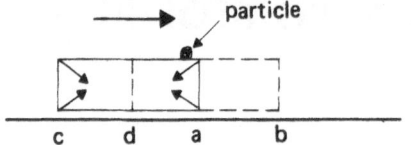

Fig. 4.1. Surface turnover as the mechanism of cell movement. New surface is produced at the front of the cell and surface resorption takes place at the back (small arrows). As the cell moves through half its own length, half of its surface is replaced. A particle on the surface remains stationary relative to the substratum.

behaviour reflects the behaviour of the cell surface. The work of Bell and Goldacre was done with free-living amoebae and that of Shaffer with the cellular slime mould. However, the suggestion that new surface may be produced during pseudopodal production has recently been made for embryonic chick heart and mouse muscle fibroblasts, again based on the observation of particles stuck to the surface [8].

It seems probable that particle behaviour may not be a reliable guide to cell surface behaviour, for two reasons. Firstly, other workers have observed that particles move forward with the cell rather than remaining stationary [36]. Secondly, a more reliable technique of surface marking, labelling with fluorescent antibody, suggests that the surface moves forward with the cell and is not renewed each time the cell moves through its own length. This has been shown both with amoebae and slime mould cells [152, 45].

Another suggestion is that the membrane may undergo active changes in shape [67]. There is no direct evidence for this idea, but a similar suggestion has been made in relation to a possible mechanism for the control of cell adhesion [72, 73]. It has been further suggested that an actomyosin-like protein at the cell

34

surface may be involved in membrane shape changes [75].

4.1.2. Haptotaxis – movement driven by interfacial forces.

The idea that cell movement may be driven by interfacial forces is an old one, but it has been plausibly restated by Carter [23]. The suggestion is that, although metabolic energy may be expended indirectly, the actual movement of a cell over a substratum is a passive phenomenon, being driven by the forces of adhesion between cell and substratum.

This idea is perhaps best understood by analogy with the behaviour of a water drop on different surfaces. When such a drop is placed on clean glass it spreads over the surface, but when placed on a 'non-stick' surface, such as Teflon, it remains approximately spherical. The drop spreads on glass because its constituent molecules adhere to the glass surface more strongly than they cohere to each other. The water molecules do not stick strongly to Teflon, so the cohesion between them maintains the drop in a spherical shape. (The cohesion between molecules is responsible for the surface tension of liquids). The adhesion between water molecules and the glass surface provides the driving force for the spreading, or movement, of the drop.

Carter found that cells would not adhere to a film of cellulose acetate, but would adhere if the film was coated with a thin layer of palladium. If the palladium was deposited in the form of a gradient on the cellulose acetate surface, cells moved up the palladium gradient and therefore up a gradient of adhesiveness. Further, cells moving over a glass surface (to which they adhere) stopped moving when they reached the boundary between glass and a cellulose acetate film. If a cell was placed on a surface composed of alternate strips of cellulose acetate and glass, the leading edge of the cell advanced over the glass strips but not over the cellulose acetate strips, resulting in a scalloped appearance (Fig. 4.2.). The suggestion made from these results

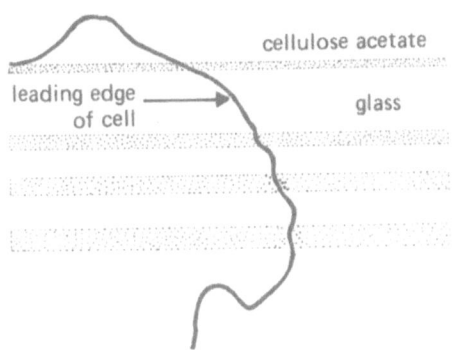

Fig. 4.2. The leading edge of a cell on a surface composed of alternate glass and cellulose acetate strips. The edge advances over the glass but not over the cellulose acetate, resulting in a scalloped appearance. (Drawn from Carter [23].)

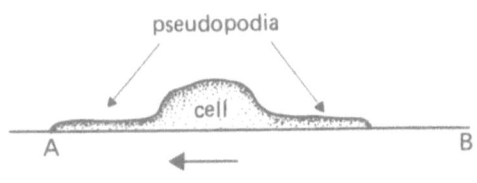

Fig. 4.3. Cell with two pseudopodia extended. Adhesion of one pseudopodium at A stronger than that of the other at B. Contraction of the pseudopodia would result in movement in the direction of the arrow.

was that the forward movement of cells is driven by the adhesive forces between cell and substratum.

It is difficult to test this idea experimentally by means of observations such as those outlined above, because we might expect a cell to move up a gradient of adhesives irrespective of how the force for movement is produced. Imagine a cell on a substratum putting out two pseudopods in opposite directions (Fig. 4.3.). Let us suppose

that pseudopod A adheres to the substratum more stongly than pseudopod B, and let us further suppose that the pseudopods are contractile (see Section 4.1.3.). Such a cell would move in direction A, because, under the force of contraction of the pseudopods, the adhesion of B would break before that of A because it is weaker. In this case the cell's movement would be *guided* by its adhesiveness to the substratum and not *driven* by it, but it would be impossible for the observer to decide which mechanism was operative.

4.1.3. Generation of the motile force by structural elements in the cytoplasm.
There are good grounds for believing that the force for cell movement is generated, at least partially, in the cytoplasm. It has been shown that cytoplasm isolated from amoebae is contractile [153] and also that this cytoplasm contains filaments, between 4 and 12 nm (1 nanometre = 10^{-9} metre = 10Å) in diameter, which may provide the structural basis for contractility [88].

In the case of cells in embryos and in tissue culture, two types of cytoplasmic filaments have been found, which possibly have specific roles in cell motility. The first type of filaments are known as *microtubules*. In longitudinal section, under the electron microscope, these appear as two parallel dark lines with a light line between, while in transverse section they appear as dark circles with light centres. They are rather more than 20 nm in diameter. Where cells are elongated, microtubules are commonly found aligned in the direction of elongation, for example, in BHK cells in tissue culture [53]. Microtubules can be disorganized by various substances such as colchicine. Cells treated with this generally cannot elongate: for example, colchicine-treated BHK cells have the rounded appearance of epithelial cells rather than the elongate, fibroblastic appearance of untreated cells [53]. It may be suggested that microtubules

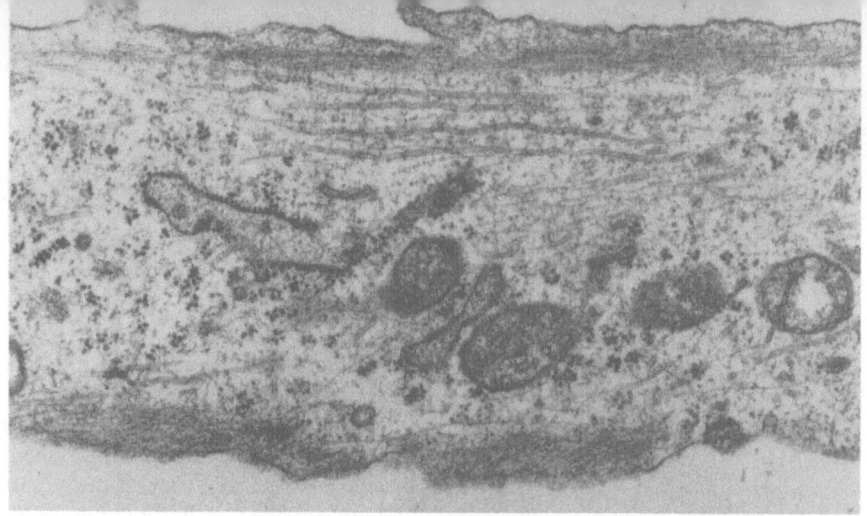

Fig. 4.4. A longitudinal section through an elongated BHK-21 fibroblast. Microfilaments are present near the upper and lower surfaces of the cell and microtubules towards the centre. Photograph kindly supplied by Dr Robert D. Goldman. From Goldman [53].)

are involved in the elongation of cells and the extension of pseudopods, and/or in the maintenance of an elongate shape. There are a number of cases in which microtubules appear to be involved in the elongation of cells during morphogenesis (see next Chapter).

The other important cytoplasmic filaments are known as *microfilaments.* Under the electron microscope these appear as solid dark lines about 5 nm. in diameter. It has been suggested that microfilaments may be involved in pseudopodal contraction, [133] and in other contractile phenomena in cells (see next chapter). The substance cytochalasin B which Carter discovered and found to inhibit cell movement, [24] disorganizes microtubules [120]. An exciting possibility is that cytochalasin B is a specific inhibitor of cell motility, but Bluemink [15] feels that its primary effect may be on the cell surface and thus on cellular adhesiveness, rather than on microfilaments. Microfilaments appear to be rather like muscle actin [71] and on this basis, Wolpert [149] estimates that 10 to 100 such filaments would be sufficient to generate the force for cell movement.

Although evidence in favour of force generation in the cytoplasm is accumulating, it is difficult to assess the contribution to motility made by the cell membrane and haptotaxis. Neither of the latter can be eliminated as possible sources of motile force. In particular, it seems certain that adhesive forces are probably important in guiding cell movement, if not, at least partially, in driving it.

4.2. Types of pseudopodal activity.

Two types of pseudopodal activity which are shown by certain embryonic as well as non-embryonic cells have been fairly extensively studied.

Primary and secondary mesenchyme cells of sea urchin embryos (see next chapter) produce long fine pseudopodia known as *filopodia.* The activity of these filopodia can be observed directly in time-lapse films of gastrulation in the sea urchin because the embryos are transparent, enabling one to see inside the blastocoele where the activity takes place. What seems to happen is that filopodia are extended from the mesenchyme cells and adhere to the inner surfaces of

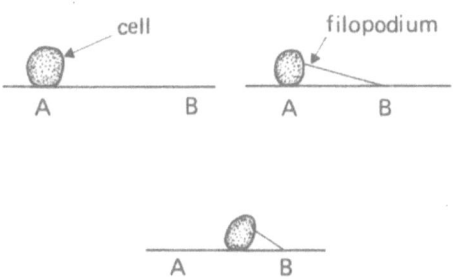

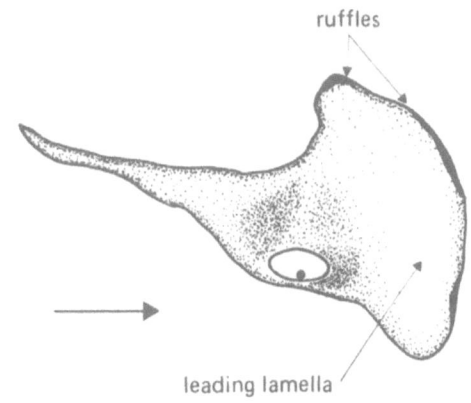

ruffles

leading lamella

Fig. 4.5. Diagram to show filopodal movement. The filopodium is extended and attaches to the substratum at B. As the cell moves from A to B, the filopodium shortens. (After Gustafson and Wolpert [60].)

the ectoderm cells which surround the blastocoele. Then the filopodia shorten as the cells are drawn forward towards the attachment points (Fig. 4.5.). The observations suggest that contraction of the filopodia moves the cells forward [60, 61]. Electron microscope observations of the filopodia of sea urchin mesenchyme cells, and treatment of gastrulae with agents which affect microtubules, suggests that microtubules may be involved in filopod extension, and microfilaments in their contraction [133]. Shortening of fine pseudopodia leading to cell movement has also been observed in certain free-living amoebae [146].

Another important type of pseudopodal activity is that shown by many embryonic and tissue cells in tissue culture, and studied most fully in embryonic chick heart fibroblasts. When moving over a substratum, these cells are characteristically roughly triangular in shape, the base of the triangle being the leading edge of the cell, (Fig. 4.6.). Under the phase contrast microscope, the front half of the cell appears devoid of cytoplasmic inclusions and this leading region has been aptly called the 'leading lamella' [70]

Fig. 4.6. Drawing of a chick heart fibroblast moving over a substratum in the direction of the arrow, showing characteristic appearance described in text.

because it is very thin and flat. Previously (and sometimes still) the leading part of the cell was called the 'ruffled membrane' because it shows 'ruffling' activity. That is to say, dark, thin, irregular lines called 'ruffles' appear at or near the leading edge of the cell. Ruffles generally last for less than one minute and during this time move backwards relative to the substratum. They may also move backwards relative to the leading lamella though usually they remain at the front edge [7].

There are a number of reasons for supposing that the leading lamella and, perhaps, ruffles are involved in the movement of the cells; not the least is the fact that they are at present at the leading edges of advancing cells (see [1]). An interesting and plausible hypothesis of how forward movement may occur has been suggested by Ingram [70], from experiments in which moving fibroblasts were viewed from the side. The suggestion is that the front edge of the leading lamella extends forward a short distance,

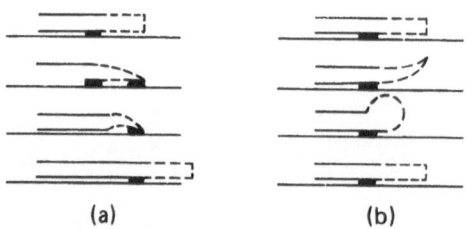

Fig. 4.7. Ingram's hypothesis of fibroblast movement and ruffling. (a) The leading edge extends, attaches to the substratum and contracts, drawing the cell forward a short distance. The cycle is repeated. (b) After extension, the leading edge fails to adhere to the substratum. Contraction causes it to bend upwards resulting in the appearance of a ruffle. (After Ingram [70].)

attaches to the substratum, and then contracts drawing the cell forward (Fig. 4.7a.). A number of such cycles of activity would obviously be necessary to move the cell forward through its own length. This idea is very similar in principle to what has been suggested regarding the activity of filopodia. Ingram further suggests that ruffles may be produced when the front edge fails to attach to the substratum and when contraction takes place towards the upper surface of the leading lamella (Fig. 4.7b.). This would cause the front edge of the cell to bend upwards and backwards, resulting in the appearance of a ruffle in a cell viewed from above. If this is so, ruffling would seem to represent a failure of the motile mechanism due to failure of adhesion with the substratum, rather than a process which is directly involved in cell movement.

Abercrombie, Heaysman and Pegrum [6, 7] have studied in detail the behaviour of the leading edges of moving fibroblasts and their ruffling activity, looking at them from above. Considering a point on the leading edge, they find a repetative sequence of advance and withdrawal covering a distance of about $5\mu m$. Points $6\mu m$ apart on the leading edge frequently behave

independently of each other: for example, while one point advances, a point $6\mu m$ away may be withdrawing. The leading edge has a net forward movement because more time is spent advancing than withdrawing. Abercrombie *et. al.* suggest that the front edge of the leading lamella is not firmly attached to the substratum at any stage [6] and show that ruffles sometimes originate *behind* the leading edge [7]. These are important points of disagreement with Ingram's hypothesis. It must be concluded that the mechanism of fibroblast movement remains unclear.

4.3. Controlling factors in cell movement.

A number of factors have been identified which seem to control, or affect, cell movement and, therefore, cell behaviour. Some of the more important of these will now be listed and discussed. Before doing this, however, it is worth stressing again that cell movements during morphogenesis are co-ordinated both spatially and temporally. Certain hypotheses have been propounded in attempts to explain the spatial co-ordination of morphogenetic movements and we may optimistically feel that we know a little about this aspect of the problem. So far as I know, practically nothing is understood about temporal co-ordination. A question which might be asked, for example, is how does the amphibian blastula 'know' when to begin gastrulation? Gastrulation may commence a certain time after the initiation of development, or when cleavage has resulted in a blastula with a certain number of cells [30], but this poses the questions of how the time is measured or the cell number counted. The answers to these questions need not be particularly complex, but I am not aware of any attempt having been made to find them. Accordingly, all examples given here will relate to the spatial co-ordination of cell movement.

4.3.1. Contact inhibition.

Contact inhibition refers to the suggestion that, when it is moving over a substratum, 'movement of a cell in the direction of a contact tends to be prevented' [5]. The idea is most easily understood in terms of the observations of Abercrombie and Ambrose [2] on moving fibroblasts. They found that when the leading lamella of a fibroblast made contact with the surface of another fibroblast, the line of demarcation between the two cells became obscured. Within a few minutes, the ruffling activity of the leading lamella usually ceased. Then, another leading lamella formed from a part of the cell surface free from contact with the other cell, and the first cell moved off in another direction.

A consequence of this behaviour is that fibroblasts tend to remain in a monolayer (i.e. a single layer of cells) on a substratum and not to climb on top of each other or move over each other. The phenomenon of contact inhibition was first recognized, in fact, from the results of studies on the monolayering of fibroblasts [4]. When a fragment of embryonic chick heart is explanted onto a suitable surface in tissue culture, fibroblasts move out from its edge over the substratum in a sheet which expands radially with uniform speed in all directions. If two such explants were placed a short distance apart, the expanding sheets of fibroblasts came into contact in the intermediate region. When this happened, the speed of movement of cells in this region was reduced compared with that of cells at the non-confronted perimeter of the expanding sheets in other regions. Following contact, cell movement in the intermediate region became randomized as opposed to radially outwards. The cells in the intermediate region did not pile up on top of each other, but remained in monolayer as judged by the lack of overlap of their nuclei.

It has further been shown that the speed of movement of individual fibroblasts is inversely proportional to the number of other cells with which they are in contact (contact number) [3]. The average speed of cells with no contacts was about 75 μm/h, whereas cells with a higher contact number (5 or 6) moved at about half this speed. Recently, it has been shown that mouse 3T3 fibroblasts exhibit a similar inverse relationship between speed and contact number, and further, that cell speed in inversely proportional to the percentage of cell perimeter which is in contact with other cells [84]. Cells with 75 to 100% of their perimeter in contact with others moved at about half the speed of cells with no contacts.

Two important questions arise from contact inhibition: (i) What is its mechanism? and (ii) Does it apply only to cells on a substratum, or, is it relevant to cells within solid tissues, as, for example, in embryos?

With regard to mechanism, there are essentially two alternative possibilities [1]. The first is that the inhibition of movement of a cell in the direction of contact is due to paralysis of the cells motile mechanism in the region of contact. This is suggested by the observed cessation of ruffling in the region of contact. Alternatively, if the cells were more adhesive to the substratum than to each other, they would be unable to leave the substratum to move over each other and, in consequence, would be unable to continue movement in the direction of mutual contact. If movement were driven by adhesive forces (see previous section) these two possibilities would seem to be equivalent. In practice, it is impossible to distinguish between them at present, since plausible but inconclusive arguments can be made in favour of each.

There is some experimental evidence relating to the relevance of contact inhibition to cells in tissues. Weston and Abercrombie [144] found no movement of cells across the boundary between fused fragments of embryonic chick heart, the cells in one fragment being labelled with ^{3}H-thymidine. In their view, this result was to be expected as a consequence of contact inhibition. However, there are cases in which cells clearly do move within solid aggregates, for example during tissue-specific sorting out (see

Section 4.3.4.). Thus, the situation with regard to the relevance of contact inhibition to the behaviour of cells elsewhere than on a substratum is paradoxical.

In conclusion, it should be noted that the term 'contact inhibition' is frequently misused. Many so-called 'normal' cell lines, when allowed to grow on a substratum, cease growth when they have formed a complete (confluent) monolayer over that surface. This growth behaviour has been termed 'contact inhibition of growth' because the state which is achieved (a monolayer) is similar to that which results from 'contact inhibition of movement'. So far as I am aware, there is no evidence that cessation of growth is a contact-mediated phenomenon. It is, therefore, more aptly described by a term which does not carry this mechanistic implication [125, 84]. 'Contact inhibition' should be used only in relation to the movement of cells, in situations where the inhibition of movement has been shown to occur as a result of contact.

4.3.2. Contact paralysis.

It has been observed in a number of cases that where cells are in contact with others they do not show pseudopodal activity. The term 'contact paralysis' has been suggested to describe this observation [61]. It implies that the motile apparatus of cells becomes paralysed locally in the region of intercellular contact. An example is the cessation of ruffling activity by fibroblasts in the region of contact formation with another cell. Another, is in the movement of epithelial sheets described by Vaughan and Trinkaus [137]. As such a sheet spreads over the substratum, the only cells which show pseudopodal activity are those at the edge, which have part of their surfaces free from intercellular contact. Cells remote from the edge of the sheet, in general do not show pseudopodal activity unless they break contact with a neighbour, in which case they may become active at the free part of their surface.

4.3.3. Contact guidance.

This term refers to the tendency for fibroblast movement to be orientated in relation to some physical orientation of the substratum. For example, they become orientated along glass fibres or shallow grooves cut into a glass surface [143]. Rosenberg [100] has shown that orientation takes place when grooves as shallow as 6 nm are present on a surface produced by building up monolayers of stearic acid on quartz. The mechanism of this orientation is not understood. However, orientation of cell movements during morphogenesis could be controlled by some orientation, for example of protein fibres, in the substratum which they move over.

4.3.4. Cellular adhesiveness.

Some ways in which cellular adhesiveness may affect cell behaviour have already been mentioned. In this section, suggestions as to how adhesiveness affects the behaviour of groups of cells and cell populations will be considered. First, however, it is important to consider the precise meaning of adhesiveness because this item is often used rather loosely. Strictly speaking, the adhesiveness of two surfaces to each other refers to the total strength or energy of adhesion between them. The total strength of an adhesion is equal to the area of the surface involved in adhesion multiplied by the energy of adhesion per unit area. A helpful way of envisaging what is meant by 'energy of adhesion per unit area' is to imagine that the surfaces are held together by chemical bonds of a certain energy. There will be a certain number of such bonds per unit area of adhering surface so that the energy of adhesion per unit area will depend upon the number of bonds per unit area and the energy of the bonds. However, this should not be taken to mean that the surfaces of adhering cells are held together by direct chemical bonds. There is, in fact, much controversy about the mechanism of cell adhesion, which is beyond

40

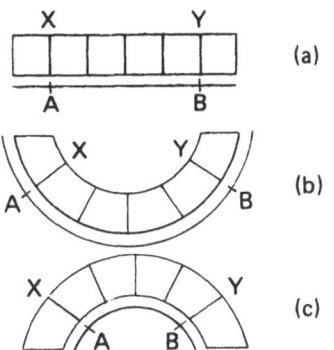

Fig. 4.8. Bending of a cell sheet brought about by changes in adhesiveness of the cells. (a) The flat sheet. (b) Bending resulting from an increase in mutual cell adhesiveness. (c) Bending resulting from a decrease in mutual cell adhesiveness. (After Gustafson and Wolpert [61].)

the scope of this book (see Curtis [33] for discussion and references).

A suggestion as to how changes in adhesiveness between cells might be important in certain morphogenetic events has been made by Gustafson and Wolpert [60, 61]. These authors were concerned with possible ways of changing the curvature of cell sheets, particularly in relation to morphogenesis in sea urchin embryos (see next chapter). Imagine a flat sheet of cuboidal cells which are adherent to each other laterally and very firmly adherent to a membrane on one side (Fig. 4.8a.). Points X and Y are on the upper surface of the sheet and A and B are on the membrane. The distances XY and AB are equal. Further, the cross-sectional areas of the cells are equal. If the mutual adhesiveness (cohesiveness) of the cells were increased, the area of adhesion between them would increase and it would do so at the expense of their free surfaces, since they are very firmly attached to the membrane on the other side. The cells would become tapered towards the top and the distance

XY would decrease, while AB remained constant. The sheet would become curved (Fig. 4.8b.). By a similar argument, but this time assuming that a decrease in cellular adhesiveness occurs, it is possible to see how curvature of the sheet in the opposite direction could occur (Fig. 4.8c.). Similar changes in the shapes of cells, but brought about by active contraction of the cells at their upper or lower ends, could bring about similar changes in the curvature of sheets. Such localized contractions may be important in certain morphogenetic events (see next chapter).

Another way in which adhesiveness may control cell behaviour has been suggested in relation to tissue-specific sorting out of embryonic cells. If cells of two embryonic chick tissues, say limb bud and liver, are dissociated and then randomly intermixed in an aggregate, they sort out according to tissue type. The two tissues adopt a sphere-within-a-sphere arrangement with cells of one type surrounding those of the other, (Fig. 4.9a, b.). In the case of liver and limb bud, the liver cells take up the outer position, surrounding the limb bud cells [121]. The sphere-within-a-sphere arrangement is an equilibrium configuration in that it can also be approached from a different starting configuration, (Fig. 4.9b, c.). Fusion of a liver aggregate with a limb bud aggregate is followed by envelopment of limb bud by liver. Certain embryonic chick tissues can be arranged in a hierarchy according to their sorting-out behaviour, [122]. The heirarchy is: back epidermis; limb bud precartilage; pigmented epithelium of eye; heart ventricle; neural tube; liver. In this series, each tissue has been found to envelop those preceeding it and to be enveloped by all those following it.

Steinberg's differential adhesion hypothesis [121, 122] attributes sorting-out behaviour to differences in the strength of intercellular adhesions. The suggestion is that cells rearrange to minimize their total adhesive free energy,

41

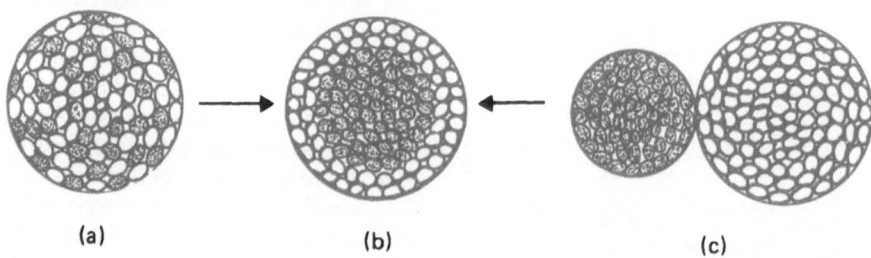

Fig. 4.9. Tissue specific sorting-out of embryonic cells. (b) shows the equilibrium sphere-within-a-sphere arrangement which can be achieved from either of the starting configurations shown in (a) and (c). Homotypic adhesions are between black cells and black cells, and white cells and white cells. Heterotypic adhesions are between black and white cells. (Partly after Steinberg [121].)

i.e. so that the strengths of their adhesions with other cells is maximal given certain geometric limitations. Thus, in a mixture of any two cell types, it is suggested that the more cohesive population sorts out internally to the less cohesive. In the case of liver and limb bud, the latter are seen as the more cohesive, and the hierarchy is regarded as a hierarchy of cohesiveness.

The sphere-within-a-sphere arrangement requires that the heterotypic adhesions (adhesion between cells of different type) should be intermediate in strength between the two types of homotypic adhesion, or at least not weaker than the weaker homotypic adhesion. This condition may be expressed as follows:

$$\left(\frac{W_a + W_b}{2}\right) > W_{ab} \geqslant W_b,$$

where W_a = strength of stronger homotypic adhesions; W_b = strength of weaker homotypic adhesions; W_{ab} = strength of heterotypic adhesions [121]. If W_{ab} were less than W_b, separation of the two intermixed cell types would result, such as occurs in species – specific sorting out of sponge cells [89].

The differential adhesion hypothesis is not the only suggested explanation of sorting-out.

All the various theories are discussed by Curtis [33].

A crucial test of any suggestion that the behaviour of cells may be explained in terms of their adhesiveness would be the direct measurement of cellular adhesiveness. In practice such measurements are extremely difficult to make. Equilibrium measurements of the cohesiveness of cells within aggregates have been made recently by Phillips and Steinberg [96]. Their results suggest that the cohesiveness of cells in this situation is, in fact, what would be predicted from the differential adhesion hypothesis. The cohesiveness of the three tissues used appears to be as follows: limb bud > heart > liver (compare with heirarchy). However, using a technique in which cells were allowed to aggregate in a linear shear gradient, Curtis [34] obtained values for cohesiveness which were the reverse of those obtained by Phillips and Steinberg. Since Curtis was dealing with the formation of adhesions, whereas Phillips and Steinberg were concerned with the strengths of adhesions which had already formed, it is perhaps not surprising that the results obtained were different, and it is possible that the latter are more meaningful in relation to sorting-out.

Roth and Weston [102] and Roth [101] obtained results which they felt suggested specific

42

adhesion between cells rather than quantitative differences in the strength of adhesions as proposed in the differential adhesion hypothesis. They used aggregates to collect single cells from cell suspensions. It was found that aggregates invariably collected more cells from the suspensions of cells of their own tissue than from suspensions of other types of cells. Superficially, this result appears conflicting with Steinberg's view that heterotypic adhesions should be of intermediate strength. However, here again the results apply to the formation of adhesions and, further, Curtis [35] could not repeat these results, arguing against the occurrence of specific adhesions between chick embryo cells.

5 Specific examples of morphogenesis

5.1. Cleavage.

Cleavage is the process of cell division by which the egg becomes divided into blastomeres. The first cleavage of the egg, which divides it into two cells, is the one which has been studied most. First cleavage in the amphibian is vertical and usually divides the egg along the future antero-posterior axis of the embryo. An indentation furrow forms at the animal pole and gradually extends around and through to the vegetal pole, thus dividing the egg into two (Fig. 5.1.).

There are a number of problems associated with cleavage, of which two will be considered here. The first and most obvious is that of the mechanism by which the furrow forms and divides the egg. The second is that division of the egg into two roughly hemispherical blastomeres necessitates the formation of 25–50% more cell surface area, so where does the new surface area come from? These problems have long been the subject of research and speculation, but recent electron microscopy seems to suggest possible answers. The latest work to be published on amphibian cleavage is that of Selman and Perry [110] and Bluemink [16]. These authors give brief reviews of and references to earlier work.

In amphibia, sea urchins, [108] annelids [10] and coelenterates [128] a ring of microfilaments (see Chapter 4) has been found immediately beneath the cell membrane at the base of the early cleavage furrow. (Fig. 5.2a.). It has been suggested that contraction of these filaments causes

Animal Pole

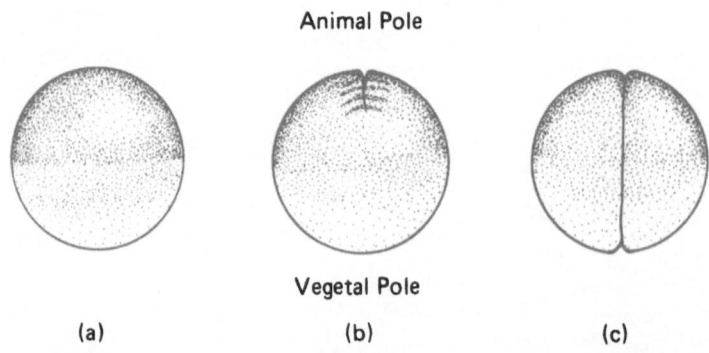

Vegetal Pole

(a)   (b)   (c)

Fig. 5.1. Cleavage of the amphibian egg. (a) Fertilized, uncleaved egg showing, from top to bottom, the pigmented animal region, the grey crescent and the white, yolky vegetal region. (b) Early cleavage furrow at the animal pole. (c) Late cleavage when the furrow has extended entirely round the egg.

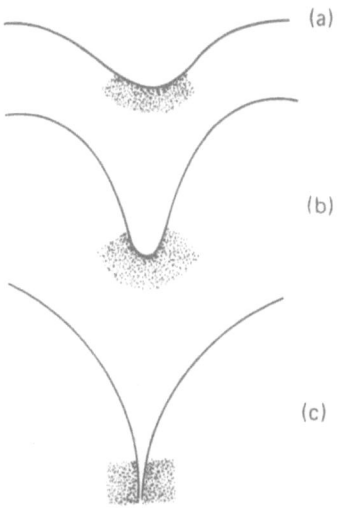

Fig. 5.2. Diagramatic sections through the cleavage furrow in the animal region, at right angles to the plane of cleavage. (a) Early furrow showing location of microfilament ring beneath the surface membrane. (b) Half way through cleavage. (c) Advanced stage of the furrow, when the filament ring has split and become orientated parallel to the cleavage plane (After Bluemink [16].)

the constriction of the cleavage furrow. In the amphibian, only the initial stages of furrow formation are thought to be produced by the filament ring [16]. Just after the stage when the furrow has passed half way through the egg, the advancing tip of the furrow seems to pass through the filament ring, splitting it into two. In the early stages of cleavage the filament ring is aligned at right angles to the cleavage plane, but after splitting the two parts of the ring become aligned parallel to the cleavage plane beneath the membrane on opposite sides of the furrow (Fig. 5.2c.).

New membrane is thought to be incorporated at the base of the cleavage furrow because the membrane in the furrow has a characteristic

appearance different from that at the egg surface [110]. After the splitting of the filament ring, progress of the furrow is thought to occur solely by a process of membrane growth. As this process takes place, point adhesions are formed between protuberances on the facing blastomeric surfaces [16]. Thus, there are thought to be two phases of furrow formation, the first brought about by a contractile ring of microfilaments, and the second by membrane growth and adhesion formation.

Treatment of eggs with cytochalasin B (see Chapter 4) did not prevent the contractile phase, nor does it disrupt microfilaments. Rather, the effect of this substance seemed to be on the formation of inter-blastomeric adhesions and membrane growth. Furrow regression took place in cytochalasin B because adhesions failed to form and membrane growth seemed to be promoted [15].

It has been shown that cortical contraction in *Xenopus* eggs can be induced by iontophoretic injection of calcium ions just beneath the surface membrane with a glass micro-electrode [51]. Contraction can also be induced by localized application of highly charged polycations, such as poly-L-lysine, to the egg surface, provided that calcium, barium or strontium ions are present in the external medium. While not discounting the possibility that polycations damage the surface membrane, Gingell suggests that they absorb to the surface altering the surface potential of the egg. The change in surface potential may lead to a change in membrane permeability allowing the influx of calcium ions which in turn cause contraction of the cytoplasm beneath the surface. The membrane surface potential is seen as a transducer, communicating environmental stimuli to the interior of the egg cell. Merely stroking the egg surface with a glass needle also induced contraction at the surface [154]. It is possible that these contractile phenomena are similar to those involved in cleavage.

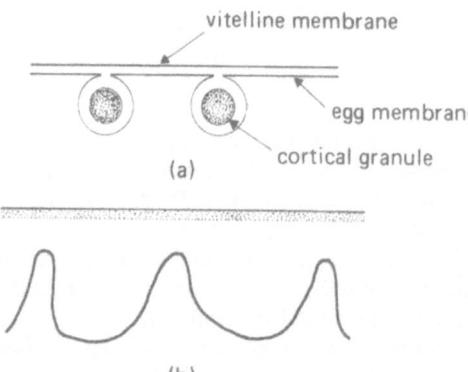

(a)

(b)

Fig. 5.3. Diagram showing how discharge of cortical granules may increase the area of surface membrane. (a) The situation before fertilization. (b) After granule discharge. The material from the cortical granules becomes attached to the inner surface of the vitelline membrane which is elevated from the egg surface. (After Curtis [33].)

Another point of interest is that the microfilaments of the contractile ring appear to be actin-like, since they form complexes with muscle heavy meromyosin [90] (see also Section 4).

It seems possible that the source of new surface membrane during cleavage of the eggs of the annelid *Pomatoceros triqueter* is different from that in amphibians [10]. It was found that the surface membrane was pleated before first cleavage, but after cleavage the pleats had disappeared, so that unfolding of the pleats could account for the increase in surface area. It was estimated that pleat unfolding could give an overall increase in surface area of 20 to 39%. The possibility that some membrane is added in the cleavage furrow was thought unlikely. It was suggested that membrane pleats are derived from the membranes of cortical granules (Fig. 5.3.). These are present beneath the surface membrane of the egg before fertilization. After fertilization they discharge their

46

contents to the outside of the egg into the perivitelline space; that is the space between the egg membrane and the surrounding vitelline membrane. It is probable that discharge of cortical granules also increases the surface area of sea urchin eggs [151]. In the newt, membrane folds present before cleavage, persist after cleavage [110].

5.2. The cellular slime mould.

During its life cycle, *Dictyostelium discoideum* undergoes a series of morphogenetic changes leading to the formation of a fruiting body in which the cellular pattern is expressed (see Chapter 2), (Fig. 5.4.). The first part of the cycle is a feeding phase in which the cells consume bacteria or other micro-organisms and multiply by division. When the food supply is exhausted, the second phase of the life cycle begins. The cells first accumulate in loose clusters. Then a dramatic process of chemotactic aggregation begins, during which the cells move in streams towards aggregation centres. In the centre, the cells pile up to form a mound which becomes surrounded by a thin, mucopolysaccharide slime sheath. From this stage onwards the cell mass, known as the grex or pseudoplasmodium, behaves as an organism in its own right, exhibiting concerted reactions to environmental stimuli such as light, temperature and humidity. First, the grex elongates away from the substratum. The elongate standing grex becomes a migrating grex by lowering itself onto the substratum and moving over it. When migration ceases, the tip of the grex moves into a vertical position, and the rest of the cell mass rounds up. Fruiting body formation then begins. The tip cells begin to form a stalk which descends through the mass to make contact with the substratum. As stalk formation continues, the cell mass is raised above the substratum and those cells which do not become part of the stalk eventually form spores.

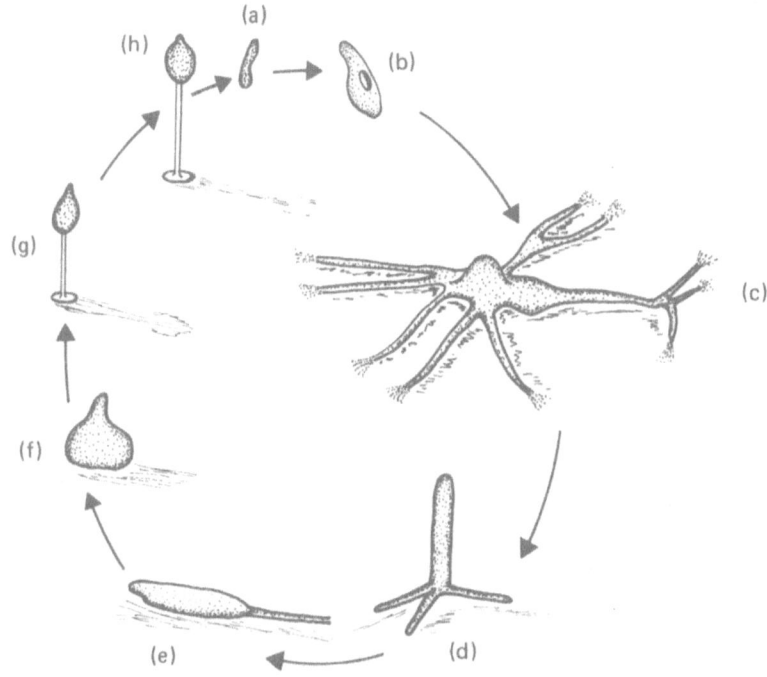

Fig. 5.4. The life cycle of the cellular slime mould, *Dictyostelium discoideum*. (a) spore; (b) free living amoeboid cell of feeding stage; (c) aggregation centre surrounded by aggregation streams; (d) standing grex; (e) migrating grex; (f) early culumination stage; (g) mid culumination; (h) fruiting body.

The whole of this life cycle is of interest in relation to morphogenesis, but one stage in particular, aggregation, has been very fully studied. One reason for this is that aggregation lends itself to microscopical observation because the cells remain adherent to the substratum. Once they have entered the aggregation centre, it becomes difficult or impossible to observe the behaviour of individual cells. There are two important aspects of the aggregation process, chemotaxis and contact interaction of the cells. Here we shall be concerned with the latter, while

chemotaxis will be considered in the next chapter which deals with intercellular communication.

Observation of the behaviour of cells on an agar surface suggests that they do not adhere strongly to each other during the feeding stage. Shortly after feeding, however, they begin to cluster together on the surface, suggesting that their mutual adhesiveness has increased [113]. Recently, it has been shown that these suggestions are probably correct. It has been found that feeding cells do not adhere to each other very much, but shortly after the cessation of

47

feeding and a considerable time before the onset of aggregation, a dramatic increase in adhesiveness takes place [43]. It is probable that this increased adhesiveness may account for the tendency of the cells to form clusters, but that a further change in adhesive properties is necessary for aggregation. This is suggested by the observation that the adhesion of post-feeding cells is inhibited by ethylenediaminetetraacetic acid (EDTA), whereas that of cells which are ready to aggregate is not [48]. (EDTA is a substance which chelates divalent cations and which is commonly found to inhibit cell adhesion or to disaggregate adherent cells.)

The adhesive properties of slime mould cells are interesting, in relation to the morphogenesis of the organism, for two particular reasons. Firstly, adhesiveness is responsible in part for keeping the later multicellular stages intact. Secondly, adhesiveness seems to be important in guiding the movement of cells towards the aggregation centre once they have joined an aggregation stream, and also in later stages of the life cycle. Cells in aggregation streams are elongated in the direction in which they are moving and adhere to each other nose to tail. In this way they follow each other into the aggregation centre. Their behaviour has been aptly called 'contact following' [114]. A cell which approaches an aggregation stream at right angles to the general direction of movement does not enter the stream as soon as it makes contact with it (Fig. 5.5.), but waits until the back of one of the stream cells passes it. It adheres to the back end of the cell, noses its way into the stream, and follows towards the centre [116]. Although it is generally agreed that contact following is important, there has been some debate about its mechanism [117, 42] which is unresolved as yet. An important point is that since contact following is not inhibited by EDTA, it is probably the EDTA-insensitive adhesions of aggregation-competent cells, which are involved in end to end contact. Possibly

48

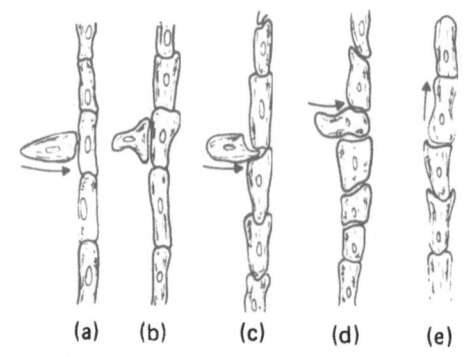

(a)   (b)   (c)   (d)   (e)

Fig. 5.5. Contact following during aggregation and the behaviour of a cell approaching and entering an aggregation stream. The direction of movement of stream cells is towards the top. (After Shaffer [116].)

these adhesions are localized to a circumferential band on the cell surface [48].

The movement of the migrating slime mould grex also presents some interesting problems, not the least of which is how its elongate, slug-like shape is generated and maintained. The grex is surrounded by a slime sheath which remains stationary relative to the substratum as the cell mass moves through it. This sheath collapses onto the substratum as the grex moves forward so that a slime trail is left behind as a record of the path the grex has taken. Because of this, the sheath must be continually synthesized, probably at the grex tip. The grex can be made to stop moving and begin fruiting body formation by disturbing the extreme tip with a fine glass needle. If the grex is cut transversely into two, the tip may continue migration but the back half commences fruiting body formation (see Chapter 3). When a number of tips are grafted laterally into a migrating grex, each is followed by the cells immediately posterior to it, so that the grex becomes divided into several smaller ones [97] (Fig. 5.6.). However, the direction of

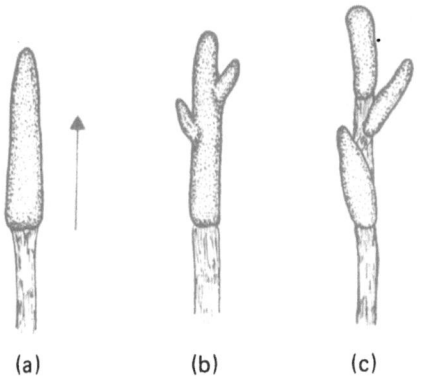

(a)　　　　　(b)　　　　　(c)

Fig. 5.6. Effect of grafting tips laterally into a
migrating grex. (a) Grex moving in direction of
arrow with slime sheath behind; (b) Additional
tips grafted on; (c) Division of the grex into
three, the cells following the tip immediately in
front of them. (After Raper [97].)

movement of a significant number of grex cells
cannot be reversed by grafting a tip onto the
back end of another grex. Nevertheless, it seems
probable that the tip and slime sheath play an
important part in polarizing grex movement.
A suggestion as to how this might occur, based
on observations of the behaviour of grex cells,
is that the slime sheath may channel the movement of the cells inside it by permitting them
to move most easily in the forward direction
[42]. This could happen if the slime sheath were
synthesized only at the tip or were more deform-
able there than elsewhere on the grex surface.
Some indirect evidence that one or both of
these possibilities may be the case has been
found [44]. Contact following is probably also
important in the polarity of grex movement.
It can be observed when a grex is squashed between a thin layer of agar and a coverslip in
order to produce a monolayer of grex cells
which can be observed microscopically.

In the migrating grex, cells are piled on top
of each other several layers deep. A nice problem

is presented in trying to understand how those
at the top move at the same speed as those on
the bottom, [114, 117]. Consider two layers
of cells on top of each other, both moving in
the same direction. If those on top use the backs
of those beneath as a substratum, they will
move forward relative to the true substratum at
twice the speed of those in the lower layer.
However, if the top and bottom surfaces of cells
were stationary during movement, as would be
the case if the cell surface turns over (see Chapter 4), the two layers could move forward at the
same speed. An alternative view is that the top
and bottom surfaces are able to slide relative to
each other and that the cells obtain traction,
not on those beneath, but on the back surface
of the cell in front [42, 149].

There are numerous other problems in relation to slime mould morphogenesis, which
cannot be discussed here. Full reviews have been
written by Shaffer [114] and Bonner [18].
Many of the experiments which are described
or referred to in these reviews can be quite
easily performed by students in the laboratory,
with the aid of simple glass tools and a binocu-
lar microscope.

5.3. Sea urchin gastrulation.
The morphogenesis of the sea urchin embryo is
probably more fully understood than any other
developmental process. One reason for this is
that the embryo is transparent, enabling the activity of cells inside it to be observed. Time-
lapse cinematography has been used extensively
to record and analyse the behaviour of these
cells. In order to use this technique effectively
it is first necessary to immobilize the embryo
which spins rapidly in the early stages, propelled
by the beating of cilia on the outer surface. One
way to accomplish this is to trap embryos or
larvae in a fine nylon net with calcium carbon-
ate crystals on the meshes [58]. Cellular aspects
of sea urchin morphogenesis have been reviewed
in detail by Gustafson and Wolpert [61]. Here

49

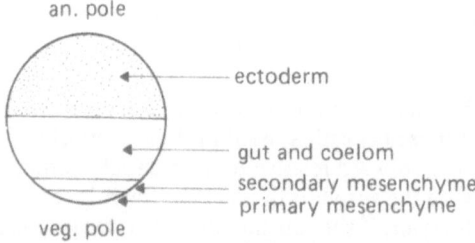

an. pole

ectoderm

gut and coelom
secondary mesenchyme
primary mesenchyme

veg. pole

Fig. 5.7. Simplified fate map of sea urchin blastula showing what has to be achieved during gastrulation. The whole vegetal half moves inside while the animal half, the ectoderm, spreads to surround it. (After Gustafson [57].)

we shall be concerned with just one stage of development, gastrualtion.

The sea urchin blastula is a hollow ball of cells, derived from the egg by cleavage. The animal half of the blastula consists of prospective ectoderm which will form the outer layer of the pluteus larva (Fig. 5.7.). (This larva is a dispersal phase only and the actual sea urchin develops later from a small part of the larva known as the 'echinus rudiment'.) The cells of the vegetal half will form the internal structures of the pluteus (gut, coelom, skeleton, etc.), but first, during gastrulation, they must move inside the ectoderm. Gastrulation is essentially a two stage process. The first stage is the inward migration of the primary mesenchyme, the cells at the extreme vegetal pole of the blastula (Fig. 5.8a,b,c.). These cells form the skeleton of the pluteus. The second stage is gastrulation proper, involving invagination of the whole vegetal half of the embryo to form the gut of the pluteus, from which the coelom arises by budding. (Fig. 5.8d,e.).

Migration of the primary mesenchyme begins when these cells lose contact with each other and with the surrounding hyaline membrane (Fig. 5.8b.). They then move into the blastocoele by an unknown mechanism and accumulate

50

at the vegetal wall (Fig. 5.8c.). Next, they migrate on the inner surface to form a characteristic pattern, a ring around the vegetal region with two forward projections (Fig. 5.8e.). This stage of their development is accomplished by means of filopodia (see Chapter 4) which are about $0 \cdot 5\,\mu m$ in diameter and up to $40\,\mu m$ long. The direction of filopod extension is random as is the initial movement of the cells. However, it is suggested that the filopodia form stronger adhesions in certain regions, and that the pattern of their final arrangement is determined by the pattern of adhesiveness of the blastocoele wall. Primary mesenchyme cells accumulate where the ectoderm is thickest and where a characteristic type of adhesion between ectodermal cells is found. In the thick regions, but not elsewhere, the ectodermal cells have inner and outer attachment points at which they adhere to each other (Fig. 5.9.). Between these attachment points, their membranes are widely separate. The filopodia of primary mesenchyme cells have been observed to adhere to the inner attachment points. Where inner attachment points are not present, the filopodia do not adhere [56]. Thus, the distribution of inner attachment sites on the inner blastocoele wall is thought to determine the distribution of primary mesenchyme cells by providing strong adhesions for their filopodia and thus directing their movement (see Chapter 4, Section 4).

The characteristic pattern of adhesion between ectoderm cells and the inner and outer attachment points with inter-membrane gaps between, may be brought about by stretching of the ectoderm. Considerable ectoderm stretching must obviously take place during gastrulation, because the ectoderm which occupies the animal region of the blastula must extend to surround the whole embryo as the cells at the vegetal region invaginate. The ectoderm is a single layer of cells so that the only way it can expand is for each cell to occupy a larger proportion of the embryo surface. A general

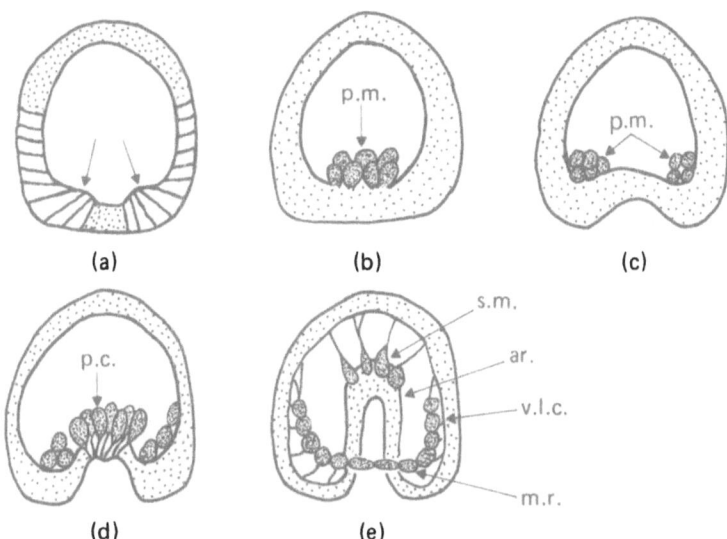

Fig. 5.8. Stages in gastrulation of the sea urchin. (a) Late blastula showing ventrolateral thickenings (arrows) and increased curvature of cell sheet. (b) Early gastrulation. Loss of adhesion of primary mesenchyme (p.m.). (c) First stage in migration of primary mesenchyme. (d) Pear shaped cells (p.c.) at archeneron tip; the first stage in invagination. (e) Late gastrula showing filopodal activity of secondary mesenchyme cells (s.m.) at tip of archeneron (ar), and pattern of primary mesenchyme which has formed ventro-lateral chains (v.l.c.) and a ring (m.r.) around the archenteron. (All after Gustafson and Wolpert [61].)

thinning of the ectoderm is observed during gastrulation.

The second phase of gastrulation, invagination of the vegetal pole material, itself appears to be divisible into two phases. This was indicated by measuring the length of the archeneron at equal time intervals throughout invagination [59]. A slow phase of inward bending of the vegetal area was followed, after a lag, by a more rapid phase of invagination.

Inward bending of the vegetal region can take place in isolated vegetal halves [87]. Because of this, it seems unlikely that invagination takes place due to a difference in hydrostatic pressure across the blastula wall. Rather, the force for invagination is generated in the vegetal region itself. It seems possible that changes in the adhesiveness of cells in the vegetal region may bring about the initial invagination by causing bending of the cell as suggested by Gustafson and Wolpert [61] (see Chapter 4). Cells in the centre of the vegetal area decrease contact with each other but, unlike the primary mesenchyme, not with the hyaline membrane. In consequence, they become pear-shaped (Fig. 5.9d.). (Note the similarity with the 'bottle-cells' in amphibian gastrulation: Section 5.4). This could result in inward bending of the vegetal region (see Section 4.3). During the late blastula stage an increase in cellular contact takes place in a circular region slightly above the vegetal pole (Fig. 5.8a.). The cells in this

51

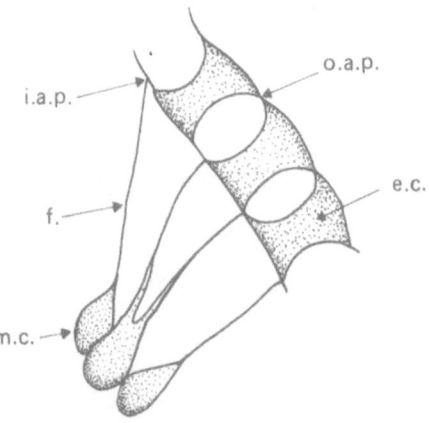

Fig. 5.9. Diagram to show outer and inner attachment points of ectodermal cells and adhesion of mesnechymal filopodia to the latter. e.c. = ectoderm cell; m.c. = mesenchymal cell; o.a.p. = outer attachment point; i.a.p. = inner attachment point; f = filopodium. (After Gustafson [57].)

region do not lose contact with the hyaline membrane, so that they become thickened and triangular in shape, with the apexes of the triangles towards the blastocoele. The curvature of the blastocoele wall increases in this region, probably because of this increase in adhesion, and the extreme vegetal pole becomes flattened. Gustafson and Wolpert [61] suggest that this ring of thickened cells limits the region of invagination during early stages of gastrulation.

The second, more rapid, stage of invagination begins when the cells at the tip of the archenteron (the future secondary mesenchyme), begin to form filopodia [37] (Fig. 5.9e.). It is suggested that the contraction of filopodia attached to the inner wall of the blastocoele provides the force for invagination. Gustafson and Wolpert [61] calculate that the force required to produce this invagination would be of the order of 10^{-2} dynes. It is probable that cells can generate

forces of this order of magnitude [149].

During the second phase of invagination, the archenteron tip moves towards the animal pole and then turns towards the ventral side of the embryo where it contacts the future mouth region. Again, it seems possible that this movement is partly directed by the distribution of ectoderm inner contact points to which the filopodia adhere.

Tilney and Gibbons [133] have shown microtubules (see Chapter 4) aligned with the long axis of filopodia of sea urchin mesenchyme cells. When colchicine and high hydrostatic pressure were applied to embryos, the cells withdrew filopodia and became spherical: gastrulation ceased. The microtubules were disrupted by this treatment. D_2O, which stabilizes microtubules, also stopped gastrulation. It was suggested that microtubules are important in determining the form of mesenchymal filopodia.

5.4. Invagination movements in amphibian development: gastrulation and neurulation.

As with cleavage, the most recent work on invagination movements in amphibia has been done with the electron microscope. The results suggest some similarities between gastrulation and neurulation movements, and possibly with the initial invagination movements of sea urchin gastrulation. We shall reverse the normal chronological sequence by dealing with neurulation first, because it appears to be the more straightforward of the two. The problem in neurulation is that of how the neural plate which is essentially flat at the end of gastrulation rolls up to form the hollow tube which will become the central nervous system (Fig. 5.10.). Baker and Schroeder [12] and Schroeder [109] have investigated the changes in shape of cells in the neural plate during rolling. The cells in the midregion of the plate are cuboidal in shape at the beginning of neurulation. As the tube rolls, they become narrow at their outer ends and elongate downwards. The narrowing of their

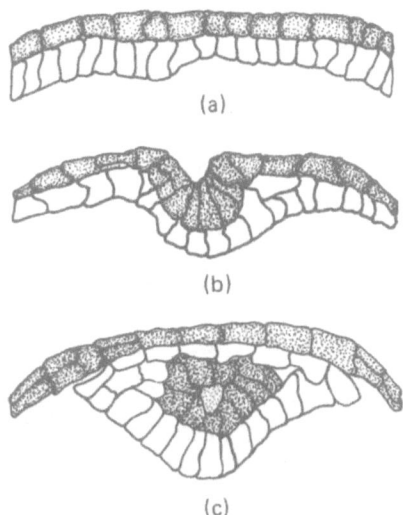

(a)

(b)

(c)

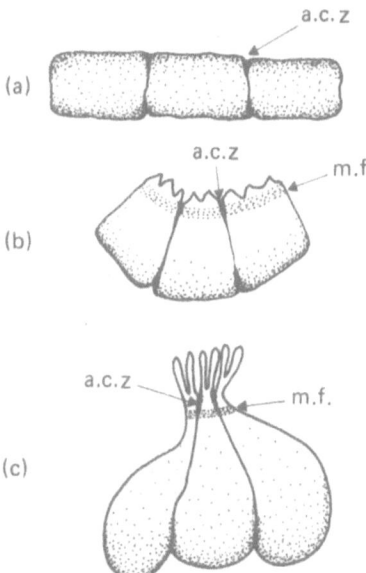

(a)

(b)

(c)

Fig. 5.10. Stages in the closure of the neural tube in amphibian showing shape changes in outer cell layer (shaded). Diagrammatic transverse sections of (a) neural plate; (b) neural groove; (c) neural tube. (After Karfunkel [74].)

Fig. 5.11. Changes in shape of cells at the centre of the neural plate showing position of microfilament ring (m.f.) and apical contact zone (a.c.z.). (a) neural plate stage. (b) neural groove stage. (c) neural tube stage. (After Baker and Schroeder [12].)

outer ends seems to be due to contraction in a plane parallel to their outer surfaces, as is suggested by the observation that their outer surfaces become thrown into folds. The contraction may be brought about by microfilaments arranged in a ring at the outer ends of these cells, parallel to and just beneath their outer surface membranes (Fig. 5.11.).

Contraction of cells on one side of a sheet could bring about sheet curvature on the side where the contraction takes place (see Section 4.3.4.), provided that the cells are firmly adherent either to an overlying membrane or laterally to each other. Baker and Schroeder show that there are desmosomes in a region which they call the apical contact zone, that is between the lateral surfaces of neural plate cells adjacent to the outer surface of the plate. Desmosomes

are specialized intercellular junctions thought to be regions of strong intercellular adhesion (Fig. 5.12).

As well as narrowing in the outer region, neural plate cells elongate during neurulation, in a direction perpendicular to the outer surface of the plate. Waddington and Perry [139] found microtubules in these cells, aligned in the direction of elongation. These microtubules may play an important role in the shape changes of the cells.

Karfunkel [74] has treated *Xenopus* neurulae with vinblastine sulphate which disrupts both microtubules and microfilaments. This treatment prevented both cell shape changes in the neural plate and neurulation. It also caused

53

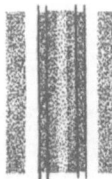

Fig. 5.12. Diagram of desmosome as seen under electron microscope. Note dense material in intercellular space and on cytoplasmic side of unit membranes.

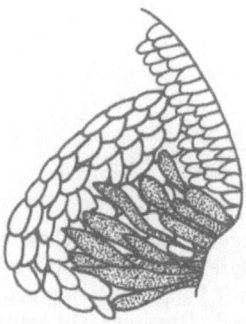

Fig. 5.13. Diagram showing 'bottle cells' (shaded) at dorsal lip of early gastrula. (After Holtfreter [65].)

reversal of cell shape changes when applied after neurulation had begun. This vinblastine sulphate treatment tends to support ideas about the suggested role of microfilaments and microtubules in causing shape changes in cells of the neural plate.

A longitudinal section through the early amphibian gastrulae reveals the presence, at the dorsal lip of the blastopore, of so-called 'bottle cells' or 'flask cells' (Fig. 5.13.). These have extremely long, thin 'necks' which are attached to the outer surface of the embryo and are orientated inwards. The inner ends of

the cells are bulbous so that the cells appear to be roughly bottle of flask-shaped. Because of their appearance at the beginning of gastrulation and their presence at the dorsal lip of the blastopore where the invagination movements begin, it seems obvious that the bottle cells have something to do with the invagination process. Bottle cells persist for some time at the tip of the archenteron as invagination proceeds, maintaining their elongation and their attachment to the inner surface of the embryo. The ventral lip of the blastopore also has bottle cells which appear later than those at the dorsal lip. There has been a considerable amount of work on amphibian gastrulation, that of Holtfreter [64, 65, 66] being particularly interesting to the advanced student. Here, however, we shall be concerned mainly with recent electron microscope observation of the bottle cells of the dorsal lip.

There appear to be certain distinct similarities between bottle cells and the cells of the neural plate during neurulation. Before gastrulation the future bottle cells are roughly cuboidal in shape, transition to their shape during gastrulation occuring during appearance of the dorsal lip of the blastopore. This shape change again involves the folding of the outer cell surfaces [11, 91] suggesting contraction in the outer region. Baker suggests that the contraction may be active and brought about by filaments in the cytoplasm adjacent to the outer surface membrane. Perry and Waddington feel, however, that the narrowing of the cell necks may occur as a result of elongation in which aligned microtubules appear to be important.

Again provided that the cells are firmly adherent, narrowing in the outer region could bring about inward bending of the gastrula surface. Holtfreter [64] suggested that there was a surface coat on the amphibian embryo, to which the cells were firmly adherent. This suggestion has been disputed though not disproved, and it seems likely that the necks of bottle cells may be firmly adherent laterally, held together, not by desmosomes, but by inter-digitating processes. Such processes are clearly seen in transverse sections through the necks [91].

Even if the narrowing of the necks of bottle cells is responsible for the initial stages of the invagination process in amphibia, it seems im-probable that it can account for later stages of gastrulation movements. The later stages of gas-trulation will not be considered here, but it seems possible to suggest that, as in the sea urchin, gastrulation in amphibia may be a two stage process, the first stage being a bending of the outer embryonic surface brought about by changes in cell shape at the blastopore lip.

It is important to mention suggestions that cellular adhesiveness may have an important function in governing the arrangement of the germ layers in the amphibian gastrula. The sor-ting-out behaviour (see Section 4.3.4.) of am-phibian embryo cells has been investigated by Townes and Holtfreter [134] and some inter-esting suggestions about the adhesiveness of the cells put forward by Steinberg [121]. Steinberg suggests that ectoderm cells have the lowest total cohesiveness (they take up the external position with respect to other tissue both in the embryo and sorting-out experiments), but that their total cohesiveness is an average of the very low cohesiveness of their (possibly coated) ex-ternal surfaces and the very high cohesiveness of their inner surfaces. Mesoderm cells may have the highest total cohesiveness because they sort out internally to both the ectoderm and endo-derm, and because they sink into the endoderm if the gastrula ectoderm is removed [66]. The position of the mesoderm in the gastrula may, therefore, be determined by its strong adhesion to the internal surfaces of ectodermal cells. The endoderm, which is less cohesive than the meso-derm (judging from sorting-out experiments), may take up the internal position because (a) it is more cohesive than the ectoderm, but (b) it cannot compete with the mesoderm for adhesion to the internal ectodermal surface.

This brief survey of morphogenesis leaves out a number of examples which are important and have received detailed study. For more de-tailed considerations you are referred to the reviews by Curtis [33] and Trinkaus [135], the latter being recommended for its account of Trinkaus' own work on *Fundulus* development.

6 Intercellular communication

It seems clear that the various pattern forming and morphogenetic processes referred to in previous chapters, since they involve cellular interactions, necessitate some form of communication between individual cells. In order to participate in these processes, a cell must 'know' something about the behaviour of its neighbours, so that its own behaviour may be co-ordinated with their's. Some suggested mechanisms for intercellular communication have already been mentioned, particularly when they have been put forward in relation to specific processes. Here we shall consider three more which may be of general importance.

6.1. Intercellular junctions of low electrical resistance.

Cell surfaces are generally of fairly high electrical resistance and of low permeability to solute molecules, providing a selective barrier between the interior of the cell and the surrounding medium. However, it has been known for some time that certain cells in the nervous system have junctions with other cells which are of relatively low resistance and allow the spread of electric current from one cell to the next. More recently, it has been shown that such low resistance junctions also exist between many non-excitable cells in embryos and in tissue culture [40].

Detection of low resistance junctions requires sophisticated electrophysiological techniques. A glass microelectrode is used to pass a current pulse into a cell and the voltage drop across the junctional membrane is recorded with another electrode in an adjacent cell. In a number of systems such measurements have indicated junctional resistances which are low compared with the resistance of the non-junctional cell surface, i.e. the electrical resistance between cells is lower than between the interior of a cell and the extracellular medium.

The implication of low resistance junctions between cells is that they are a region of high permeability through which molecules may pass from cell to cell. Passage of molecules between cells could represent a means of intercellular communication. There is some debate about the size of molecules which can pass these junctions. Furshpan and Potter [40] report that dyes with molecular weights up to 1000 pass fairly rapidly between cells. (This can be tested by injecting dyes iontophoretically into a cell and observing their spread to neighbouring cells.) In at least one case, however, that of the early *Xenopus* embryo, the dye fluorescein (molecular weight 332) has been found not to spread from one cell to another, even though the junctional membranes are of low resistance [118].

Low resistance appears to be associated with structural specialization of junctional membranes. For example, in invertebrates, particularly *Drosophila* salivary gland, [81], the intercellular junctions which have low resistance also have the characteristic ladder-like appearance known as a septate desmosome (Fig. 6.1.), a type of junction which is commonly found in invertebrates. A recent report suggests that low

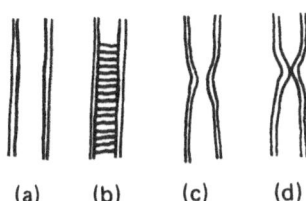

(a)   (b)   (c)   (d)

Fig. 6.1. Intercellular junctions. (a) 20 nm separation of outer leaflets of apposed plasma membranes commonly found between adhering cells. (b) Septate desmosome. (c) 'Gap' junction. Localized region, where outer leaflets approach to within 2 nm. (d) 'Tight' junction. No detectable gap between outer leaflets.

resistance junctions between vertebrate cells may occur where so-called 'gap' junctions (Fig. 6.1.) are present [49], that is where the outer leaflets of apposed plasma membranes are separated by a gap of about 2 to 3 nm. Previously, it was thought that low resistance junctions might correspond to 'tight' junctions (Fig. 6.1.), where the outer leaflets touch or fuse. Gilula et al. also demonstrate a functional significance of low resistance junctions.

6.2. Surface potential as a transducer in cellular interaction.

Due to the presence of ionogenic molecules on the outer surfaces of their membranes, cells have an electrostatic surface potential. In Chapter 5 we saw how alteration of this potential in the

Xenopus egg may bring about contraction of the cortical cytoplasm, thus acting as a transducer and means of communication between the cytoplasm of the egg and the surrounding medium. Theoretical work has suggested that surface potential could play a role in intercellular communication [50]. Calculations suggest that the close approach of cell surfaces might result in a rise in their surface potentials. Changes in surface potential might, in turn, cause changes in ionic fluxes across the membrane or molecular configurational changes at the cytoplasmic side of the membrane, which could indicate to the cell environmental changes such as the approach of another cell.

6.3. Chemotaxis.

Chemotaxis — chemically guided cell movement — differs from the above types of intercellular communication in that it does not require cell—cell contact, but involves signalling at a distance. Though it is of general biological importance, there are very few cases in which it has been shown to be involved in developmental processes. There are two types of chemotaxis, negative and positive. The former is shown by vegetative cells of the cellular slime mould (see Chapter 5) [104] and amphibian pigment cells [136]; the latter by certain spermatozoa during fertilization [103, 86] and by the aggregating cells of the cellular slime mould.

A new impetus has been given to research on positive chemotaxis in the slime mould by the discovery that the chemotactic agent, partially at least, is $3',5'$-cyclic AMP [76]. Even before this discovery some important properties of the chemotactic mechanism had been demonstrated or predicted [17, 114]. The chemotactic agent (known as 'acrasin') was found to be short-lived because it was enzymatically degraded [112]. The enzyme responsible, a phosphodiesterase, (which converts cyclic AMP to $5'$ AMP), has since been shown to be produced by the cells [25]. Further, it was felt that there was not a continuous gradient of acrasin declining radially from the aggregation centre (Chapter 5), but rather that a wave of secretion passed outwards from the centre. Thus the cells are thought to secrete acrasin in pulses, the cells in the centre secreting first. The effect on a cell or receiving a stimulatory acrasin pulse is (a) orientated movement towards the centre, and (b) secretion of a pulse of acrasin by the cell itself. Acrasin secretion is thought to be followed by a refractory period during which the cell cannot be induced to further secretion. In this way, a wave of acrasin secretion may be propagated from the aggregation centre outwards.

The aggregating cells of *Dictyostelium discoideum* can be seen to move towards the centre in pulsatile fashion in time lapse films, the bursts of inward movement being separated by intervals of about 5 min.

A detailed theoretical analysis of slime mould aggregation [28, 29] generally supports the suggestions of previous workers and suggests some new questions for experimentation. Biochemical aspects of the aggregation process are reviewed by Bonner [20]. Detailed analysis of aggregation in the slime mould, which now seems possible, may well yield results of general importance for chemotaxis.

References

1. Abercrombie, M. (1961), *Exp. Cell Res.,* Suppl. 8, 188.
2. Abercrombie, M. and Ambrose, E.J. (1958), *Exp. Cell Res.,* 15, 332.
3. Abercrombie, M. and Heaysman, J.E.M. (1953), *Exp. Cell Res.,* 5, 111.
4. Abercrombie, M. and Heaysman, J.E.M. (1954), *Exp. Cell Res.,* 6, 293.
5. Abercrombie, M., Heaysman, J.E.M. and Karthauser, H.M. (1957), *Exp. Cell Res.,* 13, 276.
6. Abercrombie, M., Heaysman, J.E.M. and Pegrum, S.M. (1970), *Exp. Cell Res.,* 59, 393.
7. Abercrombie, M., Heaysman, J.E.M. and Pegrum, S.M. (1970), *Exp. Cell Res.,* 60, 437.
8. Abercrombie, M., Heaysman, J.E.M. and Pegrum, S.M. (1970), *Exp. Cell Res.,* 62, 389.
9. Amprino, R. (1965), *Organogenesis,* Ed. R.L. DeHaan and H. Ursprung p. 255. Holt, Rinehart and Winston, New York.
10. ap Gwynn, I. and Jones, P.T.C. (1971), *Z. Zellforsch.,* 113, 388.
11. Baker, P.C. (1965), *J. Cell Biol.,* 24, 95.
12. Baker, P.C. and Schroeder, T.E. (1967), *Dev. Biol.,* 15, 432.
13. Balinsky, B.I. (1960), *An Introduction to Embryology,* 2nd. Edition. W.B. Saunders Company, Philadelphia and London.
14. Bell, L.G.E. (1961), *J. theoret. Biol.,* 1, 104.
15. Bluemink, J.G. (1971), *Cytobiologie* 3, 176.
16. Bluemink, J.G. (1971), *Z. Zellforsch.,* 121, 102.
17. Bonner, J.T. (1947), *J. exp. Zool.,* 106, 1.
18. Bonner, J.T. (1967), *The Cellular Slime Molds,* Princeton University Press, Princeton, New Jersey.
19. Bonner, J.T. (1970), *Proc. Nat. Acad. Sci. USA.* 65, 110.
20. Bonner, J.T. (1971), *Ann. Rev. Microbiol.,* 25, 75.
21. Bonner, J.T., Sieja, T.W. and Hall, E.M. (1971), *J. Embryol. exp. Morph.,* 25, 457.
22. Campbell, R.D. (1967), *Dev. Biol.,* 15, 487.
23. Carter, S.B. (1967), *Nature* 213, 256.
24. Carter, S.B. (1967), *Nature* 213, 261.
25. Chang, Y.Y. (1968), *Science* 161, 57.
26. Child, C.M. (1941), *Patterns and Problems of Development,* University of Chicago Press, Chicago.
27. Clarkson, S.G. (1969), *J. Embryol. exp. Morph.,* 21, 33.
28. Cohen, M.H. and Robertson, A.D.J. (1971), *J. theoret, Biol.,* 31, 101.
29. Cohen, M.H. and Robertson, A.D.J. (1971), *J. theoret. Biol.,* 31, 119.
30. Cooke, J. and Goodwin, B.C. (1971), *Lectures in Mathematics in the Life Sciences,* Vol. 3. p. 35. Amer. Math. Soc., Providence, Rhode Island.
31. Crick, F.H.C. (1970), *Nature* 225, 420.
32. Crick, F.H.C. (1971), *Symp. Soc. exp. Biol.,* 25, 429.
33. Curtis, A.S.G. (1967), *The Cell Surface: its Molecular Role in Morphogenesis,* London, Logos Press.
34. Curtis, A.S.G. (1969), *J. Embryol. exp. Morph.,* 22, 35.
35. Curtis, A.S.G. (1970), *J. Embryol. exp. Morph.,* 23, 253.

36. Czarska, L. and Grebecki, A. (1966), *Acta. Protozool.*, **4**, 201.
37. Dan, K. and Okazaki, K. (1956), *Biol. Bull. Mar. Biol. Lab. Woods Hole*, **110**, 29.
38. De Both, N.J. (1970), *Roux' Arch. Entwicklungsmech. Organ.*, **165**, 242.
39. Faber, J. (1971), *Adv. Morphogen.* **9**, 127.
40. Furshpan, E.J. and Potter, D.D. (1968), *Current Topics in Developmental Biology*, **3**, 95.
41. Gallera, J. (1971), *Adv. Morphogen.*, **9**, 149.
42. Garrod, D.R. (1969), *J. Cell Sci.*, **4**, 781.
43. Garrod, D.R. (1972), *Exp. Cell Res.*, **72**, 588.
44. Garrod, D.R., Palmer, J.F. and Wolpert, L. (1970), *J, Embryol. exp. Morph.*, **23**, 311.
45. Garrod, D.R. and Wolpert, L. (1968), *J. Cell Sci.*, **3**, 365.
46. Gaze, R.M. (1970), *The Formation of Nerve Connections*, Academic Press, London, New York.
47. Gerisch, G. (1960), *Roux' Arch Entwicklungsmech. Organ.*, **156**, 127.
48. Gerisch, G. (1968), *Current Topics in Developmental Biology*, **3**, 159.
49. Gilula, N.B., Reeves, O.R. and Steinbach, A. (1972), *Nature*, **235**, 262.
50. Gingell, D. (1967), *J. theoret. Biol.*, **17**, 451.
51. Gingell, D. (1970), *J. Embryol. exp. Morph.*, **23**, 583.
52. Goldacre, R.J. (1961), *Exp. Cell Res.*, Suppl. 8, 1.
53. Goldman, R.D. (1971), *J. Cell Biol.*, **51**, 752.
54. Goodwin, B.C. and Cohen, M.H. (1969), *J. theoret. Biol.*, **25**, 49.
55. Goss, R.J. (1961), *Adv. Morphogen.*, **1**, 103.
56. Gustafson, T. (1963), *Exp. Cell Res.*, **32**, 570.
57. Gustafson, T. (1964), *Primitive Motile Systems* Ed. R.D. Allen and N. Kamiya p. 333. Academic Press, New York and London.
58. Gustafson, T. and Kinnander, H. (1956), *Exp. Cell Res.*, **11**, 36.
59. Gustafson, T. and Wolpert, L. (1961), *Exp. Cell Res.*, **22**, 437.
60. Gustafson, T. and Wolpert, L. (1963), *Int. Rev. Cytol.*, **15**, 139.
61. Gustafson, T. and Wolpert, L. (1967), *Biol. Rev.*, **42**, 442.
62. Hampé, A. (1958), *J. Embryol. exp. Morph.*, **6**, 215.
63. Hay, E.D. (1965), *Organogenesis* Ed. R.L. DeHaan and H. Ursprung p. 315. Holt, Rinehart and Winston, New York.
64. Holtfreter, J. (1943), *J. exp. Zool.*, **93**, 251.
65. Holtfreter, J. (1943), *J. exp. Zool.*, **94**, 261.
66. Holtfreter, J. (1944), *J. exp. Zool.*, **95**, 171.
67. Holtfreter, J. (1948), *Ann. N.Y. Acad. Sci.*, **49**, 709.
68. Hörstadius, S. (1952), *J. exp. Zool.*, **120**, 421.
69. Huxley, J.S. and De Beer, G.R. (1934), *The Elements of Experimental Embryology*, Cambridge University Press, London.
70. Ingram, V.T. (1969), *Nature*, **222**, 641.
71. Ishikawa, H., Bischoff, R. and Holtzer, H. (1969), *J. Cell Biol.*, **43**, 312.
72. Jones, B.M. (1966), *Nature*, **212**, 362.
73. Jones, B.M. and Morrison, G.A. (1969), *J. Cell Sci.*, **4**, 799.
74. Karfunkel, P. (1971), *Dev. Biol.*, **25**, 30.
75. Kemp, R.B., Jones, B.M. and Groschell-Stewart, U. (1971), *J. Cell Sci.*, **9**, 103.
76. Konijn, T.H., van der Meene, J.G.C. Bonner, J.T. and Barkley, D.S. (1967), *Proc. Nat. Acad. Sci. USA.*, **58**, 1152.
77. Lawrence, P.A. (1970), *Adv. Insect. Physiol.*, **7**, 197.
78. Lawrence, P.A. (1971), *Symp. Soc. exp. Biol.*, **25**, 379.
79. Locke, M. (1959), *J. exp. Biol.*, **36**, 459.
80. Locke, M. (1966), *Adv. Morphogen.*, **6**, 33.
81. Loewenstein, W.R. (1966), *Ann. N.Y. Acad. Sci.*, **137**, 441.
82. Mangold, O. (1931), *Naturwissenschaften*, **19**, 905.
83. Marcus, W. (1962), *Arch. EntwMech. Org.*, **154**, 56.

84. Martz, E. and Steinberg, M.S. (1972), *J. Cell Physiol.*, **79**, 189.

85. Mettetal, C. (1939), *Arch. Anat. Hist. Embryol.*, **28**, 1.

86. Miller, R.L. (1963), *J. exp. Zool.*, **162**, 23.

87. Moore, A.R. and Burt, A.S. (1939), *J. exp. Zool.*, **82**, 159.

88. Morgan, J., Fyfe, D. and Wolpert, L. (1967), *Exp. Cell Res.*, **48**, 194.

89. Moscona, A.A. (1968), *Dev. Biol.*, **18**, 250.

90. Perry, M.M., John, H.A. and Thomas, N.S.T. (1971), *Exp. Cell Res.*, **65**, 249.

91. Perry, M.M. and Waddington, C.H. (1966), *J. Embryol. exp. Morph.*, **15**, 317.

92. Piatt, J. (1942), *J. exp. Zool.*, **91**, 79.

93. Piatt, J. (1952), *J. exp. Zool.*, **120**, 247.

94. Piatt, J. (1956), *J. exp. Zool.*, **131**, 173.

95. Piepho, H. (1955), *Biol. Zbl.*, **74**, 467.

96. Phillips, H.M. and Steinberg, M.S. (1969), *Proc. Nat. Acad. Sci. USA*, **64**, 121.

97. Raper, K.B. (1940), *J. Elisha Mitchell Sci. Soc.*, **56**, 241.

98. Rose, S.M. (1952), *Am. Nat.*, **86**, 337.

99. Rose, S.M. (1967), *Growth*, **31**, 149.

100. Rosenberg, M.D. (1963), *Science*, **139**, 411.

101. Roth, S.A. (1968), *Dev. Biol.*, **18**, 602.

102. Roth, S.A. and Weston, J.A. (1967), *Proc. Nat. Acad. Sci. USA*, **58**, 974.

103. Rothschild, Lord (1956), *Fertilization*, Methuen, London.

104. Samuel, E.W. (1961), *Dev. Biol.*, **3**, 317.

105. Saunders, J.W. (1948), *J. exp. Zool.*, **108**, 363.

106. Saunders, J.W. and Gasseling, M.T. (1968), *Epithelial-mesenchymal Interactions*, Ed. R. Fleischmajer and R.E. Billingham p. 78. The Williams & Wilkins Co. Baltimore, Maryland.

107. Saxen, L. and Toivonen, S. (1962), *Primary Embryonic Induction*, Logos Press, London.

108. Schroeder, T.E. (1969), *Biol. Bull. mar. biol. Lab. Woods Hole*, **137**, 413.

109. Schroeder, T.E. (1970), *J. Embryol. exp. Morph.*, **23**, 427.

110. Selman, G.G. and Perry, M.M. (1970), *J. Cell Sci.*, **6**, 207.

111. Sengel, P. (1971), *Adv. Morphogen.*, **9**, 181.

112. Shaffer, B.M. (1956), *Science*, **123**, 1172.

113. Shaffer, B.M. (1957), *Quart. J. micr. Sci.*, **98**, 377.

114. Shaffer, B.M. (1962), *Adv. Morphogen.*, **2**, 109.

115. Shaffer, B.M. (1963), *Exp. Cell Res.*, **32**, 603.

116. Shaffer, B.M. (1964), *Primitive Motile Systems*, Ed. R.D. Allen and N. Kamiya p. 387. Academic Press, New York.

117. Shaffer, B.M. (1965), *J. theoret. Biol.*, **8**, 27.

118. Slack, C. and Palmer, J.F. (1969), *Exp. Cell Res.*, **55**, 416.

119. Sperry, R.W. and Arora, H.L. (1965), *J. Embryol. exp. Morph.*, **14**, 307.

120. Spooner, B.M. and Wessels, N.K. (1970), *Proc. Nat. Acad. Sci. USA*, **66**, 360.

121. Steinberg, M.S. (1964), *Cellular Membranes in Development*, Ed. M. Locke p. 321. Academic Press, New York and London.

122. Steinberg, M.S. (1970), *J. exp. Zool.*, **173**, 395.

123. Stocum, D.L. (1968), *Dev. Biol.*, **18**, 441.

124. Stocum, D.L. (1968), *Dev. Biol.*, **18**, 457.

125. Stoker, M. and Rubin, H. (1967), *Nature*, **215**, 171.

126. Stumpf, H. (1967), *Arch. EntwMech. Org.*, **158**, 315.

127. Szekely, G. (1963), *J. Embryol. exp. Morph.*, **11**, 431.

128. Szolli, D. (1970), *J. Cell Biol.*, **44**, 192.

129. Takeuchi, I. (1969), *Nucleic Acid Metabolism, Cell Differentiation and Cancer Growth*, Ed. E.V. Cowdry and S. Seno p. 297, Pergamon Press, New York.

130. Taylor, A.C. (1943), *Anat. Rec.*, **87**, 379.

131. Thornton, C.S. and Thornton, M.T. (1965), *Experentia*, **21**, 146.

132. Tiedemann, H. (1966), *Current Topics in Developmental Biology*, **1**, 85.

133. Tilney, L.G. and Gibbons, J.R. (1969), *J. Cell Sci.*, **5**, 195.

134. Townes, P.L. and Holtfreter, J. (1955), *J. exp. Zool.*, **128**, 53.

135. Trinkaus, J.P. (1969), *Cells into Organs: The Forces that Shape the Embryo*, Prentice-Hall Inc., Englewood Cliffs, New Jersey.
136. Twitty, V.C. and Niu, M.C. (1954), *J. exp. Zool.*, **125**, 541.
137. Vaughan, R.B. and Trinkaus, J.P. (1966), *J. Cell Sci.*, **1**, 407.
138. Waddington, C.H. (1966), *Major Problems in Developmental Biology*, Ed. M. Locke p. 105. Academic Press, New York.
139. Waddington, C.H. and Perry, M.M. (1966), *Exp. Cell Res.*, **41**, 691.
140. Webster, G. (1966), *J. Embyrol. exp. Morph.*, **16**, 105.
141. Webster, G. (1966), *J. Embryol. exp. Morph.*, **16**, 123.
142. Webster, G. and Wolpert, L. (1966), *J. Embryol. exp. Morph.*, **16**, 91.
143. Weiss, P. (1958), *Int. Rev. Cytol.*, **7**, 1.
144. Weston, J.A. and Abercrombie, M. (1967), *J. exp. Zool.*, **164**, 317.
145. Wilby, O.K. and Webster, G. (1970), *J. Embryol. exp. Morph.*, **24**, 595.
146. Wohlman, A. and Allen, R.D. (1968), *J. Cell Sci.*, **3**, 103.
147. Wolpert, L. (1969), *J. theoret. Biol.*, **25**, 1.
148. Wolpert, L. (1970), *The New Scientist*, **46**, 322.
149. Wolpert, L. (1971), *The Scientific Basis of Medicine Annual Reviews*, p. 81.
150. Wolpert, L., Hicklin, J. and Hornbruch, H. (1971), *Symp. Soc. exp. Biol.*, **25**, 391.
151. Wolpert, L. and Mercer, E.H. (1961), *Exp. Cell Res.*, **22**, 45.
152. Wolpert, L. and O'Neill, C.H. (1962), *Nature*, **196**, 1261.
153. Wolpert, L., Thompson, C.H. and O'Neill, C.H. (1964), *Primitive Motile Systems*, Ed. R.D. Allen and N. Kamiya p. 143. Academic Press, New York.
154. Zotin, A.I. (1964), *J. Embryol. exp. Morph.*, **12**, 247.
155. Zwilling, E. (1961), *Adv. Morphogen.*, **1**, 301.

Index